Amobarbital Effects and
Lateralized Brain Function

David W. Loring   Kimford J. Meador
Gregory P. Lee   Don W. King

Amobarbital Effects and Lateralized Brain Function

The Wada Test

Springer Science+Business Media, LLC

David W. Loring
Section of Behavioral Neurology
Department of Neurology
Medical College of Georgia
Augusta, GA 30912-3275

Kimford J. Meador
Section of Behavioral Neurology
Department of Neurology
Medical College of Georgia
Augusta, GA 30912-3280

Gregory P. Lee
Department of Psychiatry and
  Department of Surgery (Neurosurgery)
Medical College of Georgia
Augusta, GA 30912-4010

Don W. King
Epilepsy Diagnostic and Treatment Unit
Department of Neurology
Medical College of Georgia
Augusta, GA 30912-3200

With two figures.

Library of Congress Cataloging-in-Publication Data
Amobarbital effects and lateralized brain function : the Wada test / David W. Loring
  . . . [et al.].
     p.   cm.
  Includes bibliographical references and index.
  ISBN 978-1-4612-7704-0   ISBN 978-1-4612-2874-5 (eBook)
  DOI 10.1007/978-1-4612-2874-5

  1. Wada test.   2. Amobarbital—Diagnostic use.   3. Epilepsy.
  I. Loring, David W.
  [DNLM: 1. Amobarbital—diagnostic use.   2. Neuropsychological
  Tests.   WL 103 A523]
  RC473.W34A48   1992
  616.8′075—dc20
  DNLM/DLC
  for Library of Congress   91-5160

Production managed by Hal Henglein; manufacturing supervised by Jacqui Ashri.
Camera-ready copy prepared by the authors.

9 8 7 6 5 4 3 2 1

To our families, our teachers, and our patients

Preface

The intracarotid amobarbital (or Amytal) procedure is commonly referred to as the Wada test in tribute to Juhn Wada, the physician who devised the technique and performed the first basic animal research and clinical studies with this method. Wada testing has become an integral part of the pre-operative evaluation for epilepsy surgery. Interestingly, however, Wada initially developed this method as a technique to assess language dominance in psychiatric patients in order that electroconvulsant therapy could be applied unilaterally to the non-dominant hemisphere.

Epilepsy surgery has matured as a viable treatment for intractable seizures and is no longer confined to a few major universities and medical institutes. Yet, as is increasingly clear by examining the surveys of approaches used by epilepsy surgery centers (e.g., Rausch, 1987; Snyder, Novelly, & Harris, 1990), there is not only great heterogeneity in the methods used during Wada testing to assess language and memory functions, but there also seems to be a lack of consensus regarding the theoretical assumptions, and perhaps, even the goals of this procedure.

Our book is intended to present our clinical biases regarding Wada testing, and describe research results on brain-behavior relationships that have been derived from this procedure. We do not feel that there should necessarily be a single Wada test that is employed by all centers any more than there should be a single operative technique for epilepsy surgery. Our writing of this volume should not be construed as an attempt to provide our approach as the best procedure to assess language and memory functions. What we are attempting to do is to examine critically some of the assumptions needed for this test, examine what the literature reports on the clinical findings derived from this test, in order that results of Wada testing are not interpreted uncritically, both for group descriptions as well as for individual patient inferences. The first portion of the book deals primarily with clinical issues related to Wada testing, namely Language and Memory assessment, as well as amobarbital effects on the EEG. In addition, the Wada protocol that we employ at the Medical College of Georgia is presented in the final chapter of the book.

Wada testing provides a unique opportunity to examine the effects of primarily unilateral cerebral dysfunction on cognitive and behavioral functions. Chapters 4 and 5 are less clinical in nature, since they consider amobarbital effects on both attention and emotion. Since patients are tested both prior to and following hemispheric anesthesia, each subject can serve as his or her own control, and greater understanding of brain-behavior relations can be attained. When making comparisons of left vs. right hemisphere functioning, it is important for the times at which behaviors are measured to be matched. The deficits observed following injection are transient and in continuous resolution. Without concern for this important detail, there will be a tendency for "right hemisphere behaviors" to be tested somewhat earlier given the absence of aphasia. Although knowledge of the vascular distribution of cerebral arteries is important for all chapters, this has been detailed in Chapter 5 on emotion, given the role frequently attributed to subcortical structures in emotional response.

Finally, there has recently been a tendency away from calling this procedure the Wada test and, instead, referring to it as either the intracarotid sodium amytal (ISA) test, or the intracarotid amobarbital procedure (IAP). Although the uncontrolled proliferation of acronymious medical jargon is sufficient reason to continue the use of the "Wada test," the primary reason we continue its use is out of respect for Dr. Wada. Anyone who has had the privilege to speak with Dr. Wada, about topics ranging from epilepsy to the arts, will acknowledge that this continuing honor and recognition is justifiably deserved.

July 13, 1991   *David W. Loring*

Acknowledgments

Although this volume reflects individual experiences and opinions, they could not have been derived outside the context and without support of the Medical College of Georgia Epilepsy Surgery Program. Foremost has been the contribution of Herman F. Flanigin, M.D., the founding director of our epilepsy surgery program. Dr. Flanigin performed his Neurosurgery Fellowship at the Montreal Neurological Institute under the direction of Wilder Penfield, M.D., and was the junior author on their landmark paper entitled "Surgical therapy of temporal-lobe seizures" published in 1950 in the *Archives of Neurology and Psychiatry*. Dr. Flanigin not only provided the wisdom derived from 40 years of experience in epilepsy surgery, but also was the direct link to the institution whose name is synonymous with epilepsy surgery excellence. To our knowledge, Dr. Flanigin is the only individual with sufficient expertise in multiple disciplines to individually perform temporal lobectomies in what was primarily a private practice setting.

Brian B. Gallagher, M.D., Ph.D., was another key individual, who with Dr. Don W. King, established the MCG epilepsy program prior to Dr. Flanigin's arrival. Dr. Gallagher's exceptional sensitivity to patient care further distinguishes him. Anthony M. Murro, M.D., has provided valuable consultation on hardware. Joseph R. Smith, M.D., our current epilepsy program director, was trained by Dr. Flanigin. From our perspective with specialized interest in behavioral correlates of brain function, Dr. Smith's undergraduate degree in psychology facilitates a very collegial relationship.

Departmental support from Thomas R. Swift, M.D., has been essential for the successes of our program, and his encouragement of clinical research has been much appreciated. We are also grateful to the ongoing institutional as well as state support for this program. Other individuals at MCG who deserve recognition include Debbie Cliatt, Judy Jackson, Karen Stanford, and Wanda Littleton, not only for their daily effort which results in a smoothly running program, but for their continuing help with our research studies. Pat Downs has been the major organizer for preparation of this manuscript, and our overly qualified proofreader, Cheryl A. Gratton, Ph.D., has again demonstrated her superior verbal processing skills.

Finally, we acknowledge the significant contribution made by our patients. We are grateful for their willingness to participate in many of our studies during what is frequently the most stressful period of their lives.

Contents

1
Language

Although Wada (1949) developed the technique of selective hemispheric anesthetization to identify language dominance, the first report of local anesthetic application to functionally inactivate cerebral language areas was described by Gardner (1941). Two left-handed patients were being evaluated for surgery, and Gardner was aware that the patients' handedness raised the possibility of reversed cerebral laterality. If these patients were right language dominant, Gardner reasoned, a left resection could be more extensive if performed on the non-language dominant hemisphere. Alternatively, resection involving the right hemisphere could potentially compromise language. "The removal of a tumor at the cost of the patient's speech is scarcely an accomplishment on which to congratulate oneself" (p. 1035).

Both patients received procaine hydrochloride to identify cortical language zones. A hypodermic needle was introduced through a trephine opening anterior to the right facial motor area in the first patient. Twenty-seven cc of 0.75% novocaine were injected producing a left facial paralysis without aphasia. Consequently, a right hemispherectomy was performed without post-operative aphasia. To evaluate for residual left hemisphere language functioning in the second patient, 10 cc of 1% solution of procaine hydrochloride were injected anterior to the left facial motor area without producing aphasia. In contrast to the first patient, no facial weakness was noted, which the authors interpreted to indicate the facial motor area was no longer functional. Following resection, which included Broca's area, no aphasia was present.

Gardner concluded that knowing the patients' language dominance aided in determining the degree of resection. However, the lack of aphasic disturbance in no way guarantees that language is represented in the contralateral hemisphere. In only 1 case did a functional motor impairment develop from the injection, although Gardner states that "no one has produced aphasia by injection of procaine into Broca's area, but that it would follow seems beyond doubt" (pp. 1037-1038). Gardner considered the amount of procaine used in the first case to be excessive, with cavities found at the injection site when the specimen was resected. No further reports of this technique appeared. Eight years later, Wada's first paper was published on his technique of intracarotid amobarbital injection, which has since become the standard technique to establish cerebral language dominance.

The prevalence of left-cerebral dominance has been estimated between 92% and 99% for dextral subjects (Annett, 1975; Benson & Geschwind, 1985; Hécaen, Mazars, Ramier, Goldblum & Mérienne, 1971). However, the pattern of cerebral language asymmetry is less clear in non-dextral (i.e., left- or mixed-handed) patients. Among non-dextral patients, right-hemisphere language has been estimated as high as 40% (Roberts, 1969). Gloning (1977) examined the incidence of aphasia in patients with structural lesions who eventually came to autopsy. Eighty percent of the dextral patients with left hemisphere lesions were aphasic, whereas no dextral patients with right hemisphere lesions displayed impaired language function. In contrast, the presence of aphasia was approximately 80% in non-dextral patients regardless of hemispheric involvement side. These data suggest significant language bilaterality in non-dextral patients. Bilaterality of language is also suggested by reports of more complete recovery from aphasia in non-dextral individuals than dextral patients (Gloning, 1977). Summarizing the incidence of aphasia with unilateral hemispheric lesions, Benson (1985) calculated that 60% of dextral patients with damage to the left hemisphere developed aphasia, whereas only 32% of non-dextral patients with left hemispheric damage become language impaired. Similarly, only 2% of dextral patients sustaining injury to the right hemisphere become aphasic. In contrast, 24% of non-dextral patients are aphasic following right-hemisphere injury.

Although not without limitations, bilateral intracarotid amobarbital testing provides the best current technique to directly assess each hemisphere's contribution to language function. Throughout this chapter, we will use the terms "language dominance" and "speech dominance" somewhat inter-changably. Some epilepsy surgery centers prefer the use of "speech dominance," referring to the vocal production of language. As described by Snyder, Novelly & Harris (1990), speech does not include comprehension. The barbiturate inactivation of speech areas produces aphasic errors or speech arrest which are used for determining cerebral dominance.

Our bias is to use "language" rather than "speech," although we have sometimes retained "speech" if that was the expression used in the original report. Although, strictly speaking, the use of "speech" is correct, it is similar to a behavioristic approach to explaining human behavior in which the only subject of study was behavior that could be observed, avoiding theoretical concepts that could not be observed. Philosophically, we are using the Wada test to make inferences about specific brain functions. In addition, we prefer "language" over "speech" because speech can be affected by peripheral factors, and because speech arrest may be seen following right hemisphere injection, without other evidence of right hemisphere linguistic contribution (Oxbury & Oxbury, 1984; Loring, Meador, Lee, Murro, Smith, Flanigin, Gallagher &

King, 1990). Comprehension is a fundamental aspect of language, and can be easily tested during the Wada (McGlone, 1984). Further, left hemisphere injection has been shown to affect sign-language communication (sign-language aphasia), clearly indicating the specialized role of the left hemisphere in linguistic processing independent of speech (Damasio, Bellugi, Damasio, Poizner & Van Gilder, 1986). However, we do not consider changes in prosody, which can also be considered an important language component, to be important for the clinical language evaluation since it is non-linguist and tends to be associated with right hemisphere function (Ross, Edmondson, Seibert & Homan, 1988). As with all other scientific reports, careful operational definition of the constructs under study will minimize discrepancies reported in the literature.

Studies from the Montreal Neurological Institute

Original Reports. The first English article describing amobarbital assessment to determine language representation was written by Juhn Wada and Theodore Rasmussen in 1960. Twenty patients were described who received administration of 150-200 mg amobarbital via a common carotid artery stick.

The first large-scale patient series undergoing Wada testing to assess speech and language function was reported by Branch, Milner & Rasmussen (1964). The majority of the 123 patients presented were either left-handed or ambidextrous. Some right-handed patients were also evaluated if clinical doubt existed regarding language representation. Four patients in this series were excluded due to unsatisfactory studies. Of the remaining 119 patients, bilateral studies were conducted in 97 patients; 22 patients had unilateral tests only. Surgery was conducted in 99 of the 119 patients.

The patients received a 200 mg injection while counting. Speech testing included object naming, counting, and reciting the days of the week both forwards and backwards. Of the 71 left-handed and ambidextrous patients, 34 patients, or 48%, had speech represented in the left hemisphere; 27 patients, or 38%, had language represented in the right; whereas in 10 patients, or 14%, speech was thought to be bilaterally represented. This figure contrasts with the 48 right-handed patients reported, 43 of whom were determined to be left hemisphere speech dominant. Five right-handed patients, or 10%, had right hemisphere language, with no right-handed subjects having bilateral speech.

Patients with evidence of early left cerebral injury within the first 5 years of age (criteria unspecified) displayed a different pattern. Of the 27 patients with early left injury, 6 patients (22%) had speech on the left, 18 patients (67%) had speech on the right, and 3 patients (11%) had bilateral language. This contrasted with 28/44 (64%) left cerebral dominant non-dextral patients with no evidence of early injury, 9 patients (20%) with right hemisphere speech dominance, and patients (16%) with bilateral language representation.

Confirmation of speech representation was provided by speech mapping at the time of surgery or by transient postoperative aphasia following language dominant resections. Of 44 patients in whom the surgery was performed on the language dominant hemisphere, 40 displayed language representation using electrical stimulation at the time of surgery or with postoperative aphasia. An additional 3 patients were found to have postoperative verbal impairment on psychological testing, although no aphasia, which the authors interpreted as mild evidence confirming left cerebral language representation.

In 46 patients with resection of the nondominant hemisphere in or near areas whose contralateral homologous areas are known to be language related, 45 had no postoperative language deficits. One patient, who was classified as right hemisphere language dominant, suffered a mild but definite aphasia following left hemisphere resection involving Broca's area.

These data are important for several reasons. First, they empirically demonstrate that the age of cerebral insult is an important factor in determining cerebral language lateralization. Also, they illustrate the statistical maxim that the null hypothesis cannot be proven. In 1 patient who was thought to be right hemisphere cerebral dominant and who underwent left resection, a mild post-operative language deficit was observed. On the Wada test, no language representation was noted following left hemisphere injection.

Certain patients in this series were classified solely on the basis unilateral injections. Only 97 patients received bilateral injections, whereas 22 received unilateral injections only. However, when Branch, Milner & Rasmussen (1964) classified speech lateralization as left, right, or bilateral, 119 patients are included. Therefore, some patients were classified as having language restricted to a single hemisphere based upon unilateral injections only, thereby decreasing the reported incidence of bilateral speech representation.

Milner, Branch & Rasmussen (1966). The issue of bilateral representation was addressed by Milner, Branch & Rasmussen (1966). The patient series included 18 patients with presumed bilateral language, 17 of whom were non-dextral. This number represented approximately 15% of the 117 non-dextral patients studied. Two patterns of bilateral amobarbital language impairment were described. In 6 patients, no speech arrest was noted from either side although clear evidence of aphasia was observed following each injection. In the other 12 cases, the period of muteness, which typically lasts up to 2 minutes following left hemisphere injection in right-handed patients, lasted only 30-60 seconds on both sides. Further, 9 of the 18 patients displayed an asymmetrical pattern of language impairment. In 7 patients, object naming impairment with preserved series repetition ability was present following left hemisphere injection, whereas series repetition impairment with intact confrontation naming followed right hemisphere injection. In 2 other patients, the reverse pattern of language asymmetry was noted.

Milner (1975)/Rasmussen & Milner (1975). The next major report of Wada testing was published by Milner (1975) and Rasmussen & Milner (1975). At this time, a series of 371 patients was included. In 140 right-handed patients without clinical evidence of early left hemisphere injury, presumably still defined as 5 years of age or less, 134 (96%) patients were left hemisphere language dominant whereas 6 (4%) were right cerebral dominant. No cases of bilateral speech were present in the dextral patients without evidence of early left brain injury. For the 122 non-dextral patients, 82 were determined to be left hemisphere speech dominant, with the remaining 36 patients equally divided between right and bilateral speech (18 patients each, 15% of sample). In contrast, the 109 patients with evidence of early injury demonstrated greater bilateral and right hemisphere speech representation. In 31 right-handed patients, 25 (81%) were left, 4 (13%) were right, and 2 (6%) displayed bilateral language. In the non-dextral sample of 78 patients with early left-hemisphere injury, the majority of patients 40 (51%) were right hemisphere language dominant only, 23 (30%) had speech restricted to the left hemisphere, and 15 (19%) displayed bilateral language. Thus, it appeared that both handedness and the presence of early left hemisphere injury contributed to bilateral and right hemisphere speech representation.

Rasmussen & Milner (1977). The patient sample size was further increased in the largest and most commonly cited patient series (Rasmussen & Milner, 1977). Since the 1975 publications, 25 additional patients with clinical or radiological evidence of early left-brain injury were included. Their series of patients without evidence of early left hemisphere injury was not updated, however, since the focus of the report was on the effect of early brain injury on language representation. Thus, their reported prevalence indicating that 96% of dextral patients without evidence of early damage to the left hemisphere were left hemisphere dominant for language, with all remaining patients being right hemisphere language dominant are the same values as reported in 1975. This is also true for the left- or mixed-handed patients without early injury, who exhibited less laterality to the left, with only 70% displaying left language dominance, 15% bilateral, and 15% right hemisphere language dominant.

The slightly larger sample of patients with early injury did not appreciably alter the reported frequency of speech representation, although the values are probably more reliable given the larger sample size. Of the 42 right-handed patients with early left hemisphere injury, 34 (81%) were left hemisphere speech dominant, 5 (12%) displayed right hemisphere language, and 3 (7%) were mixed dominant. Of the 92 left-handed patients with early left brain injury, 26 (28%) were left cerebral dominant, 49 (53%) were right cerebral dominant, and 17 (19%) displayed bilateral speech representation.

Montreal Conclusions. The data presented by Montreal provided an excellent introduction to language representation with and without evidence of early injury. However, they should not be interpreted uncritically and generalized without caution to other populations for several reasons. The early procedures were done without angiography using a direct carotid stick, and extent of cross-flow or abnormal vasculature could not be documented. Further, the number of patients with incomplete data, either for technical or behavioral reasons, is not indicated. As we know from Branch, Milner & Rasmussen's 1964 paper, some patients with only unilateral injections were included in the entire patient series. It seems reasonable to assume that these same patients were included in the subsequent reports that updated their experience with this technique.

The sensitivity of Wada testing was also shown to be less than perfect in the 1 case determined to be right cerebral language dominant, but who developed aphasia following left-sided surgery. This observation that patients with right cerebral language dominance may have additional left hemisphere language areas was confirmed by Wyllie, Lüders, Murphy, Morris, Dinner, Lesser, Godoy, Kotagal & Kanner (1990). These authors observed left hemisphere language representation with subdural electrode array stimulation in patients with right cerebral dominance as indicated by Wada testing. In contrast, none of the left cerebral language dominant patients displayed any evidence of right language representation from stimulation mapping. They expressed concern regarding the ability to exclude left hemisphere language on the basis of right cerebral language dominance following amobarbital injection, and suggested that all patients without exclusive left hemisphere language undergo language mapping for both extensive left and right temporal and frontal resections. These findings indicate that the sensitivity of Wada testing depends on multiple factors, including how language is assessed during the period of hemispheric anesthesia, and as we will discuss below, the criteria used for determining language impairment.

Bilateral Language

Milner, Branch & Rasmussen (1966) described 2 patterns of bilateral amobarbital language impairment, one of which included aphasia following both injections, and the other consisting of asymmetrical language impairment (L>R and R>L). The type of language assessment used during the period of hemispheric anesthesia will affect the reported frequency of language findings. Object naming and series repetition were the initial measures of language function in the Montreal series, although spelling and reading were added later. The importance of comprehensive language assessment is underscored by the 1 patient in the initial report of Branch, Milner & Rasmussen (1964) who, despite the amobarbital speech testing suggesting only right hemisphere language representation, became transiently aphasic following left hemisphere surgery. Thus, the procedures employed at that time did not appear sensitive to relatively subtle language representation and may have underestimated the frequency of the asymmetric bilateral language representation.

Oxbury & Oxbury (1984) reported a higher incidence of bilateral language representation than the Montreal series. They administered 175-200 mg amobarbital to 17 dextral and 6 non-dextral subjects. In the 23 patients who were selected for amobarbital testing due to concern for laterality of language or adequacy of memory function, 8 patients displayed aphasic errors following left hemisphere injection only, 1 patient displayed aphasic errors following right hemisphere injection only, and 14 patients presented with aphasic errors following injection to both cerebral hemispheres. These aphasic errors included deficits in naming and forward/backward series recitation (e.g., days of the week, months of the year, counting). The differences in extent of bilaterality of language suggest that Wada testing procedures and patient inclusion criteria, in addition to the relatively low base-rate occurrence of mixed- and right-hemisphere language dominance, are contributing to inconsistencies in reported language representation.

Following injection of approximately 150 mg amobarbital, Hommes & Panhuysen (1970) observed right hemisphere language impairment in 9/11 depressed patients (10 dextral, 1 non-dextral), although the impairment was much less than that following left hemisphere injections. Powell, Polkey & Canavan (1987) administered approximately 75-125 mg amobarbital to 27 patients (18 dextral, 9 non-dextral) and observed mixed cerebral dominance in 6/27 patients (22%) and right hemisphere speech in only 4 patients (15%). Overall, 29 patients were dextral and 44 were non-dextral. Zatorre (1989) reported that in his sample of 61 patients who were administered 150-175 mg (38 dextral, 23 non-dextral), only 35 (57%) were left hemisphere dominant for speech, whereas 22 (36%) were considered bilateral, and 4 (7%) were right hemisphere speech dominant.

Rey, Dellatolas, Bancaud & Talairach (1988) reported their cumulative experience with 73 patients from 1961 to 1983. As with most other centers, their procedure varied over time. The earlier patients (1961 to 1965) received medication injection via direct common carotid artery puncture with a mean dosage of 195 mg amobarbital. From 1966 to 1970, the average dosage was 87 mg and after this time, an average dose of 110 mg was administered. The patients reported by Rey, Dellatolas, Bancaud & Talairach (1988) underwent Wada testing if a left seizure onset began during childhood, if they were non-dextral, or if they presented with speech impairment during or immediately after a seizure. Twenty-nine patients were dextral and 44 were non-dextral. Language dominance was determined by counting cessation, aphasic response, comprehension impairment, repetition errors, and reading errors. Of the 29 right-handed patients, 27 (93%) were left cerebral dominant for speech, and 2 (7%) displayed bilateral language representation. In contrast, 15/30 left-handed patients (50%) were right cerebral language dominant, 11/30 (37%) were left speech dominant, with 4 patients (13%) displaying bilateral language. Of patients with mixed-handedness, 7/14 (50%) displayed left hemisphere language, 2/14 (14%) displayed right hemisphere language, and 5/14 (36%) displayed bilateral speech. No effect of medication dosage administered was observed. Fifteen patients had a right-sided hemiparesis, which would increase the likelihood of exclusive right hemisphere speech representation. Other studies with sample sizes greater than 50 are presented in Table 1-1 for comparison purposes.

Table 1-1 Incidence of language dominance from several Wada series. Note that some values have been recombined from the original articles to facilitate comparison. Data are from entire series without respect to age of probable CNS damage or to hand preference.

	Left	Bilateral	Right
Loring, Meador, Lee, Murro, et al. (1990)			
Representation	79	22	2
n=103	(77%)	(21%)	(2%)
Relative Dominance	92	5	6
	(89%)	(5%)	(6%)
Kurthen et al. (1991)			
Representation	54	21	4
n=79	(68%)	(27%)	(5%)
Relative Dominance	71	4	4
	(90%)	(5%)	(5%)
Mateer & Dodrill (1983)			
n=90	75	6	9
	(83%)	(7%)	(10%)
Rey et al. (1988)			
n=73	45	11	17
	(62%)	(15%)	(23%)
Rasmussen & Milner (1977)			
n=396	280	38	78
	(71%)	(10%)	(20%)
Rausch & Walsh (1984)			
n=62	53	4	5
	(86%)	(6%)	(8%)
Strauss & Wada (1983)			
n=78	63	5	10
	(81%)	(6%)	(13%)
Strauss et al. (1987)			
n=90	63	17	10
	(70%)	(19%)	(11%)
Woods et al. (1988)			
n=208	187	11	10
	(90%)	(5%)	(5%)
Zatorre (1989); n=61			
n=61	35	22	4
	(57%)	(36%)	(7%)

(Updated from Loring, Meador, Lee, Murro, Smith, Flanigin, Gallagher & King, 1990; reproduced with permission of *Neuropsychologia*)

Medical College of Georgia Study

We contrasted hemispheric language representation to relative hemispheric dominance using the Wada test in 103 patients (Loring, Meador, Lee, Murro, Smith, Flanigin, Gallagher & King., 1990). Patients who had greater linguistic impairment following a single injection, yet still had language representation on the contralateral side, were distinguished from those who only displayed deficits following a single hemisphere injection. Since abnormal radiologic findings in non-temporal regions reportedly affect the frequency of atypical language representation (Woods, Dodrill & Ojemann, 1988), patients with radiologic evidence of lesions outside the temporal lobe were excluded. Depending on the criteria employed, patients could be considered to be either dominant for language in a single hemisphere using relative language asymmetry criteria of dominance, or determined to have bilateral language representation if the presence of any language deficits following unilateral injection are considered.

Our first language classification (exclusive language representation) considered the presence of linguistic errors following each hemisphere independently. Language impairment following both injections led to classification of bilateral language representation regardless of relative asymmetry of impairment. With this classification technique, language deficits had to be observed following a single injection only in order to infer language dominance.

Our second classification procedure (forced relative dominance) was based upon relative language impairment. A patient displaying bilateral language impairment, but with a greater deficit following a single injection, was classified as language dominant for the hemisphere with greater linguistic failure following intracarotid amobarbital injection. Patients were considered to have bilateral language only when a side of greater representation could not be determined. To determine language asymmetry, the sum of language ratings was calculated for each side, and laterality ratios were computed (i.e., L-R/L+R). Subjects with laterality ratios greater than 0.15 or less than -0.15 were classified as having either left or right language dominance, whereas patients with laterality scores between 0.15 and -0.15 were classified as having bilateral language with no asymmetry. Thus, a patient displaying bilateral but asymmetrical language representation was classified according to the side of greater language disruption (e.g., L>R classified as L).

Seventy-nine patients (77%) had exclusive left hemisphere language representation, 2 patients had exclusive right hemisphere language representation, and 22 patients had language to varying degrees in each hemisphere. Three patterns of language representation were observed in patients with bilateral language representation: 17/22 bilateral language patients (77%) displayed asymmetric representation (13L>R, 4R>L); no relative dominance was present in the remaining 5 patients. Table 1-2 presents our results of hemispheric language representation as a function of handedness. Based upon exclusive language representation, 80% of the dextral patients displayed exclusive left hemisphere language; 19% had bilateral language, and 1 patient had language impairment following right hemisphere injection only. Of the 12 non-dextral patients, 6 had exclusive left hemisphere language, 1 patient had exclusive right hemisphere language, and 5 displayed bilateral language. When examining relative language dominance, 91% of the dextral patients were left hemisphere dominant, 4% right hemisphere dominant, and 4% mixed dominant (1 each in the latter 2 categories). Of the 12 non-dextral patients, 9 were left hemisphere dominant, 2 were right hemisphere dominant, and 1 mixed cerebral dominant. Thus, classification by relative language dominance instead of language representation decreased the number of patients with mixed cerebral dominance.

Two additional patient groupings were then formed, each of which attempted to minimize the effects of early injury on cerebral language lateralization. The initial grouping excluded those patients with evidence of brain injury prior to age 2. The injuries were typically perinatal complications, but evidence of any CNS dysfunction, including a single seizure episode, was an exclusion criterion. The grouping resulted in 86 patients without evidence of early injury, the results of which are also presented in Table 1-2. This more restricted grouping did not appear to appreciably alter the reported percentages for dextrals, and the remaining non-dextral sample size was insufficient to make any inferences. Subjects with evidence of early injury prior to age 6 were then excluded, and the frequency of language and hand dominance was recomputed. Although this grouping creates even smaller sample sizes, the results are also presented for descriptive purposes.

The frequency of non-dextral patients in the exclusive left hemisphere language groups (6/79) was contrasted to the frequency of non-dextrals in the mixed and exclusive right hemisphere language group (6/24). Thus, there was a higher proportion of non-dextrals when language was not exclusively represented in the left hemisphere.

Our technique is sensitive to relatively subtle language representation. Nevertheless, we do not feel that it is overly sensitive because intracarotid amobarbital results are confirmed by functional cortical mapping for language and because our overall estimate of 80% left hemisphere language dominance is in the same general range as other amobarbital series (Table 1-1). Thus, we believe our procedure assesses linguistic functions without overestimating the occurrence of language representation. In our experience, the presence of exclusive right hemisphere language representation, with no left hemisphere language representation, is a rare event. In our patients without evidence of early injury, a single dextral patient displayed exclusive right hemisphere language representation. Similarly, in our non-dextral group, only 1 patient displayed exclusive right hemisphere language, and he was omitted when forming the most restricted grouping without early brain injury because his seizures began at age 3 years.

Table 1-2 Language representation in MCG series as a function of handedness with different classification of early brain injury

Language Representation Without Regard to Early Injury				
Handedness	N	Left	Bilateral	Right
Exclusive Language Representation				
Right	91	73 (80.2%)	17 (18.7%)	1 (1.1%)
Left/Mix	12	6 (50.0%)	5 (41.7%)	1 (8.3%)
Forced Relative Dominance				
Right	91	83 (91.2%)	4 (4.4%)	4 (4.4%)
Left/Mix	12	9 (75.0%)	1(8.3%)	2 (16.7%)

Language Representation Without Early Injury (<2 years)				
Handedness	N	Left	Bilateral	Right
Exclusive Language Representation				
Right	77	63 (81.8%)	13 (16.9%)	1 (1.3%)
Left/Mix	9	6 (66.7%)	2 (22.2%)	1 (11.1%)
Forced Relative Dominance				
Right	77	72 (93.5%)	2 (2.6%)	3 (3.9%)
Left/Mix	9	7 (77.8%)	0 (0.0%)	2 (22.2%)

Language Representation Without Early Injury (<6 years)				
Handedness	N	Left	Bilateral	Right
Exclusive Language Representation				
Right	66	56 (84.8%)	9 (13.6%)	1 (1.5%)
Left/Mix	3	1 (33.3%)	2 (66.7%)	0 (0.0%)
Forced Relative Dominance				
Right	66	61 (92.4%)	3 (4.5%)	2 (3.0%)
Left/Mix	3	2 (66.7%)	0 (0.0%)	1(33.3%)

(From Loring, Meador, Lee, Murro, Smith, Flanigin, Gallagher & King, 1990; reproduced with permission of *Neuropsychologia*)

Similar results have been reported in the only other study that has explicitly differentiated between representation and relative dominance. Kurthen, Linke, Reuter, Hufnagel & Elger (1991) presented results of bilateral Wada studies in 79 patients, and observed complete left cerebral dominance in 54 patients, mixed language dominance in 21 patients, and 4 patients with right cerebral language. Although they did not provide the criteria for determining language dominance, they stated that in their sample of 21 mixed language dominant patients, there was relative superiority of the left hemisphere in 17 patients. Although not stated, the remaining 4 mixed language patients did not display a significant asymmetry of language representation.

Reasons for Discrepancies. As Woods, Dodrill & Ojemann (1988) note, the Rasmussen & Milner (1977) results are frequently cited without reservation as being representative of the general population with respect to language lateralization, and these authors also reported data that were discrepant from the initial report of Rasmussen & Milner. As can be seen in Table 1-1, Rasmussen & Milner reported a sizable proportion of their patients to be right hemisphere language dominant (19.7%). Only those patients in whom atypical language lateralization is suspected undergo Wada testing at the Montreal Neurological Institute, which pioneered this procedure. In a similar vein, when examining the effects of early injury on language representation, only patients with early left hemisphere injury were examined (Rasmussen & Milner, 1977).

Language determination was initially based upon interruption of counting and the presence of paraphasic responses during rote serial speech (e.g., reciting days of the week) or oral spelling. No formal assessment of comprehension was employed, and it is unclear whether confrontation naming was assessed. Further, sentence repetition was not included in their language assessment, a task that is frequently the most sensitive to subtle linguistic impairment.

With any new technique, a natural evolution of the procedure with increasing experience is expected, and such an evolution is suggested by Zatorre (1989). In his sample of 61 patients, most of whom were unilateral temporal lobe epilepsy patients, only 35 (57%) were left hemisphere dominant for speech, whereas 22 (36%) were considered bilateral and 4 (7%) were right hemisphere speech dominant. This figure of 36% of patients with bilateral language, obtained from patients at the Montreal Neurological Institute, is considerably higher than the 10% reported by Rasmussen & Milner (1977). Thus, sample inclusion characteristics as well as criteria for impaired language will affect any series' results.

Snyder, Novelly & Harris (1990) surveyed 55 different epilepsy surgery centers to examine the criteria employed for inferring bilateral language representation. Of the 47 centers reporting specific percentages, there was a range from 0% speech bilaterality to 60% bilaterality (see Table 1-3). In addition, diverse criteria for inferring language representation were reported. Most centers (93%) employed naming as a criterion for establishing speech lateralization for the dominant hemisphere. Other criteria included unspecified aphasic signs (78%), counting ability (80%), familiar word/phrase repetition (61%), and unfamiliar word/phrase repetition (65%). In contrast, heterogeneous criteria were present for inferring the presence of speech in the non-dominant hemisphere. The criteria included attempts to mouth appropriate

Table 1-3 Variability in reported percentage of patients with mixed speech dominance (MSD)

% MSD	Centers Reporting	Sample %
0%	11	20%
1-2%	2	4%
5-6%	9	17%
10%	13	24%
15%	4	7%
16-29%	7	13%
60%	1	2%
TOTAL 47 (87% of centers responding)		

(From Snyder, Novelly & Harris, 1990; reproduced with permission)

words, groaning sounds, singing ability, object naming, partial phoneme vocalization, serial rote speech, and expression of familiar words.

To test whether criteria discrepancies were accounting for differences in the reported frequency of mixed speech dominance, Snyder, Novelly & Harris subjected these data to a discriminant function analysis. Nearly 70% of centers responding were classified correctly into 1 of 2 groups (0-6% mixed language incidence vs. 10%-20%, plus an additional center reporting 60% mixed language representation) based on whether vocalization of partial phonemes, serial rote speech, and expression of familiar words were assessed to determine language representation. A lower prevalence of mixed language was associated with procedures employing serial rote speech, expression of familiar words, or production of partial phonemes as evidence of language in the hemisphere contralateral to injection. Snyder's survey also indicated that approximately 15% of the centers do not perform amobarbital testing on all surgical candidates, and instead, only on those patients with evidence of bilateral brain disease. Thus, without consecutive patient series, non-biased estimates cannot be obtained. Further, estimates will vary based upon the primary patient population being evaluated (e.g., temporal lobectomy candidates vs. hemispherectomy candidates with hemiplegia). Causality cannot be inferred based upon correlational results, since the relationship may be due to a different variable, such as whether all patients are included. The Snyder data, however, illustrate the importance of method variance when attempting to interpret the amobarbital literature.

A continuum from left hemisphere language dominance to right hemisphere language dominance should be expected on the basis of the handedness literature. Patients are frequently rated on their degree of handedness, ranging from strongly dextral, moderately dextral, mixed, to moderately sinistral and strongly sinistral. To the degree that there is a relationship between handedness and cerebral language dominance, one should expect similar degrees of mixed language dominance. That is, one should *not* expect language laterality to be primarily discrete when handedness is a continuous variable. Consequently, the previously reported incidence of right hemisphere language dominance likely reflects, in part, dichotomizing a continuous variable. In addition, the high incidence of right hemisphere language dominance may reflect the absence of comprehensive language assessment during amobarbital anesthesia. In our experience, language restricted only to the right hemisphere is relatively rare, and in the absence of purely left hemisphere language, most patients exhibit bilateral representation.

Factors Associated with Atypical Representation

There are several consistent findings in patients who develop mixed, or right cerebral language dominance, the most common of which are early brain injury and non-dextral handedness. In terms of early brain injury, the most critical factor producing a shift in language is probably age at which the injury occurred, and secondarily, what brain regions are affected by the damage. Non-dextral handedness is also frequently associated with a shift in language representation, although this relationship is probably non-causal, with both handedness and shift in language representation related to the underlying neural substrate.

Age of Injury. Rasmussen & Milner (1977) reported the first strong relationship between age of injury to the left hemisphere and language reorganization as determined by Wada testing. In patients who sustained left cerebral injury prior to 6 years of age, 34 of 42 right-handed patients, or 81%, displayed left cerebral language dominance. In contrast, only 26 of 92 non-dextral patients, or (28%), displayed left cerebral language dominance, with 17 patients presenting with bilateral language (19%) and 49 patients (53%) displaying right cerebral language dominance.

This observation has been made by others. Rey, Dellatolas, Bancaud & Talairach (1988) reported that patients with right cerebral language developed their seizures at an earlier age, and in addition, identifiable cerebral lesions were more frequent. Of particular importance is that the presence of a neurological deficit, in particular right hemiparesis, was statistically more frequent in this group. This finding agrees with Ajersch & Milner (1983) who observed atypical language representation in all 13 patients with a right hemiparesis. Among Ajersch & Milner's 12 sinstrals with right hemiparesis, 8 were right speech dominant.

Strauss & Wada (1983) observed a relationship between age of onset and language laterality. Seven of 11 patients, or 77%, with right hemisphere language had left hemisphere injury or seizure development prior to age 1. In contrast, left hemisphere patients whose injury occurred after age 1 tended to be left cerebral language dominant (19 of 22, or 86%).

Satz, Strauss, Wada & Orsini (1988) examined cerebral reorganization by reviewing studies that employed either Wada testing, dichotic listening, or post-surgical aphasia. Across techniques, the earlier the lesion occurred, the greater the likelihood that a shift in both handedness and language laterality would be observed.

Lesion Variables. Rasmussen & Milner (1977) reported that lesions sparing the primary frontal and parietal language areas did not affect language representation. Patients with radiologic evidence of left hemisphere abnormalities outside the temporal lobe more frequently displayed atypical language (10 of 29) as compared to similar patients with right lesions (1 of 18; Woods, Dodrill & Ojemann, 1988). In addition, 9 of 10 patients with left hemisphere injury of sufficient magnitude to result in right hemiparesis had atypical (i.e., right or mixed) cerebral language representation.

Handedness. The relationship between handedness and atypical language representation is unclear. Rasmussen & Milner (1987) suggested that "an early lesion that does not modify hand preference is on the whole unlikely to change speech representation" (p. 359). Thus, unless a patient were left-handed, standard cerebral language representation was considered likely. The shift from right- to left-handedness as a consequence of early left hemisphere injury has subsequently become known as pathologic left-handedness (e.g., Satz, Orsini, Saslow & Henry, 1985). However, Rausch & Walsh (1984) reported that right hemisphere language dominance can be observed in dextral patients whose seizures originate from the left temporal lobe, although this finding was not observed by Woods, Dodrill & Ojemann (1988). Again, some of these differences may be related to differences in patients studied, as well as type of language assessment and criteria employed.

Rey, Dellatolas, Bancaud & Talairach (1988) agreed with Rasmussen & Milner's observation that although shift in either handedness or language dominance from the left to the right hemisphere can occur independently as a function of early left hemisphere injury, the more likely occurrence is that both shift together. Rey, Dellatolas, Bancaud & Talairach observed that motor dominance, but not language dominance, could shift laterality, with 3 sinistral patients with right hemiparesis displaying left hemisphere language. The remaining patients with right hemiparesis presumably presented with bilateral language. In contrast, Rausch & Walsh (1984) have demonstrated that language dominance can shift independently from manual preference.

Strauss & Wada (1983) reported that injury to the left hemisphere prior to age 1 year was associated with a decline of right handedness, footedness, and eyedness. A slight but nonsignificant increase in dextrality was associated with the early right brain injury group. Of the patients with a shift in language or hand preference, Satz, Strauss, Wada & Orsini (1988) reported 68% had shifts in both language and handedness. A shift in language dominance to the right hemisphere without an accompanying shift in handedness was seen with later occurring lesions, but still occurring before age 6 years. Rey, Dellatolas, Bancaud & Talairach (1988) also discussed the possibility of natural left-handedness with right speech dominance in which a right brain lesion induces a manual shift without a language shift.

Woods, Dodrill & Ojemann (1988) compared their sample of 10 right cerebral language dominant patients to their 11 patients with bilateral speech, and found that all patients with exclusive right cerebral language representation were left-handed, whereas only 5 of 11 mixed language dominant patients were sinistral. This difference in frequency was statistically significant ($p \leq .01$).

Summary. As observed by Wood, Dodrill & Ojemann (1988), the incidence of handedness and speech lateralization using Wada testing will not be representative of the normal population and should not be generalized to it. However, across the series of studies, several consistent findings appear.

The age at which the injury occurs is the most important finding. The greatest shift in language representation is seen with injuries to the left hemisphere prior to age 1 when cerebral plasticity appears greatest. The larger the injury to the left hemisphere, the greater is the likelihood that language will shift to the right hemisphere (e.g., right hemiplegia). Although lesions outside the temporal lobe are more likely to produce a shift in laterality, temporal lobe lesions can also produce a language shift.

Handedness is a marker of shift in language laterality. A change from dextral to sinistral as a function of left brain injury, pathologic left handedness, occurs frequently. However, handedness and cerebral dominance can shift independently of each other, although it is more likely that if only a single shift occurs, it is in the language domain. With increasing markers for possible shifts in language representation (e.g., early left brain injury, left injury involving greater areas of tissue, left handedness), there is an increased likelihood of exclusive right cerebral language representation.

Language Validation

Snyder, Novelly & Harris (1990) argue for validating the Wada test prior to making any inferences regarding the prevalence of bilateral language representation. They suggest that the amobarbital procedure could be an initial first step in language identification, with utilization of "additional techniques such as Positron Emission Tomography, cortical mapping via subdural electrode grids, or Single Photon Emission Tomography. However, the sensitivity and specificity of the non-invasive procedures suggested for validation remain to be validated themselves. It is unlikely that bilateral subdural electrode grids would be frequently employed, and without bilateral grids, patients with presumed bilateral language could not be identified. Further, it is unlikely that as much cortical area could be tested using subdural grids as that assessed with carotid barbiturization. PET studies are unlikely to have either the spatial or temporal resolution to identify all specific language regions.

Branch, Milner & Rasmussen (1964) employed several techniques to validate the initial use of Wada testing. The techniques used for validation were identifying speech areas with cortical mapping and observing post-operative aphasia in patients with surgery to the speech dominant hemisphere. Although Wada testing remains primarily validated on patients whose seizures originate from the left cerebral hemisphere, we have had 3 patients with seizures originating from the right temporal lobe in whom bilateral language representation (2 L>R, 1R>L) was suggested by Wada testing. (Loring, Flanigin, Meador, Lee & Smith, 1988). During the Wada evaluation, these patients demonstrated disruption of counting/sequencing, impaired comprehension, and/or produced paraphasic errors with right intracarotid amobarbital injection. Functional cortical mapping of right hemisphere language regions in these patients revealed a combination of either posterior frontal and/or perisylvian regions that, when stimulated, produced speech arrest or paraphasic substitution during speech recitation. Similarly, Rosenbaum, DeToledo, Smith, Kramer, Stanulis, & Kennedy (1989) reported a patient with a right seizure onset and right hemisphere language inferred from Wada testing in whom right hemisphere language representation was confirmed with cortical language mapping during surgery. We also tested language following surgery in our single patient with greater right than left language representation (Loring, Meador, Lee, Flanigin, King & Smith, 1990). Significant paraphasic responses were observed in all core linguistic domains (i.e., naming, repetition, generative fluency, comprehension, reading, writing). Thus, we have convergent evidence of right hemisphere language representation in this patient, confirming the results obtained with Wada testing.

Another criticism offered by Snyder, Novelly & Harris (1990) is that the procedure assumes that all cortical areas in the hemisphere responsible for speech production possess the same physiological threshold for anesthetization both within and between patients. Further, Snyder, Novelly & Harris state that certain language areas may lie outside the regions supplied by the anterior and middle cerebral arteries. However, most areas thought to be important for language are affected by internal carotid artery injection, and further, areas that might be compromised by temporal lobectomy are supplied by the internal carotid artery. It is not necessary to have the same physiological threshold for all areas or to completely anesthetize all language zones to produce aphasia. This would be analogous to stating that we have learned nothing about aphasia unless all linguistic behavior was affected at the time of the lesions.

The only concern regarding incomplete anesthetization would be an incorrect inference of bilateral language based upon a finding of no speech disruption following injection to either side. Nearly 78% of the centers responding to the Snyder survey indicated that they require the presence of aphasic signs prior to inferring language representation. Thus, few centers appear willing to infer bilateral language representation on the basis of bilaterally normal language.

As previously discussed, the absence of linguistic impairment following amobarbital injection does not guarantee the absence of language representation in that hemisphere (e.g., Milner, Branch & Rasmussen, 1964; Wyllie, Lüders, Murphy, Morris, Dinner, Lesser, Godoy, Kotagal & Kanner, 1990). Hart, Lesser, Fisher, Schwerdt, Bryan & Gordon (1991) extended this conclusion to specific aspects of linguistic function (i.e., comprehension of word meaning). These authors presented a picture-word association task during the period of hemispheric anesthesia. The task consisted of 4 picture and word pairs, 2 of which had semantically related associations (e.g., bed [picture] and sleep [word]), and 2 of which were unrelated (e.g., bus [picture] and scale [word]). In addition to stimulus identification, the patient was to determine if the 2 items were related by category or function. Despite having difficulty reading the words or naming the pictures following left hemisphere injection, 8 of 15 patients, or 53%, were able to perfectly match the visually presented words to their corresponding pictures by their meanings and associations. Since performance on semantic-relatedness tasks appears dependent on the dominant hemisphere posterotemporal-inferoparietal region, and non-dominant hemisphere lesions do not produce semantic errors on picture-word matching, the authors concluded that the most parsimonious explanation of this sparing is that the posterotemporal-inferoparietal region may not be sufficiently impaired by amobarbital injection. Although we do not agree that using the Wada test for language evaluation requires the assumption that "carotid injection of amobarbital can reach all brain areas of one hemisphere that are involved in language" (Hart et al., 1991, p. 55), it is clear that if surgery is contemplated in this region, then additional measures should be undertaken to identify language representation.

The importance of multiple language measures during Wada testing is illustrated by one of our patients, who displayed speech arrest lasting approximately 25 seconds following right hemisphere injection. Despite this apparent speech arrest, all remaining language domains assessed were normal. Because we were uncertain whether or not speech cessation reflected language representation in the right hemisphere, functional cortical mapping for language was carried out prior to right temporal lobectomy. Despite sufficient threshold for sensorimotor mapping, no language impairment was present during electrical stimulation of either frontal or temporal regions. Further, no impairments were present during repeated language testing conducted during the first post-operative week. This patient, therefore, supports our decision to require impairment in multiple language domains prior to inferring language representation. Other authors (Oxbury & Oxbury, 1984) have similarly noted that occasionally patients will develop speech arrest following injection to the non-language-dominant hemisphere. Again, this is one reason why we prefer "language" over "speech" when referring to the linguistic capacity of the brain.

Validation of Non-Invasive Behavioral Techniques

Since the Wada test is the least ambiguous technique available to establish cerebral language representation, it has frequently been employed as the criterion against which non-invasive cerebral dominance techniques have been evaluated. The non-invasive techniques generally present competing verbal stimuli through a single sensory modality (e.g., visual, auditory), and immediately following presentation, the subject is to identify the presented stimuli (Kimura, 1961; Bryden & Rainey, 1963). In general, patients tend to more accurately identify items presented to the right side of the body, indicating the greater contribution of the left hemisphere in processing the material. Even in the auditory modality, in which there are considerable bilateral projections from the sensory organ, there is greater contralateral vs. ipsilateral projection. Thus, when any deviation is present from expected advantage for material presented to the right side, a conclusion of atypical cerebral language representation can be considered.

Strauss, Wada & Kosaka (1985) studied visual half-field effects in 41 epilepsy surgery candidates, 6 of which displayed bilateral language and 5 of which had right hemisphere speech. Four-letter words were presented bilaterally. They found that left cerebral language dominant patients tended to show a right visual half-field advantage, and patients with mixed or right cerebral language exhibited a bias in favor of the left visual half-field. However, the authors cautioned that visual half-field testing was not entirely reliable, with the predicted laterality of language dominance correct in only approximately 70% of the sample.

Strauss, Gaddes & Wada (1987) examined dichotic listening, employing a free-recall paradigm, in 90 epilepsy surgical candidates grouped according to left cerebral language (n=63), mixed cerebral language (n=17), and right cerebral language (n=10). Patients with left cerebral language tended to show the expected right ear advantage in word identification. Patients with right cerebral language showed no lateral preference, and bilateral language patients tended to show a right ear advantage. For individual patient inference, the prediction of speech representation by direction of ear advantage was 80%. Since the results were replicable, the authors concluded that "a free-recall verbal dichotic listening test provides relevant, albeit not totally accurate, information regarding cerebral speech dominance in epileptic patients" (p. 753).

Zatorre (1989) studied dichotic listening performance in 67 epilepsy surgery candidates. In contrast to previous Wada/dichotic listening studies, he employed a fused rhymed task. The stimuli consisted of rhymed consonant-vowel-consonant words that differed only on the initial consonant (e.g., "coat" presented to the left ear, "goat" presented to the right ear. The stimuli are generally perceived as unitary and localized to the midline. Thus, with the above example, a patient with a right ear advantage would respond that only "goat" was heard. In 33 of 35 left cerebral language patients, right ear advantages were obtained. Similarly, all 4 patients with right hemisphere language displayed a left ear advantage. Patients with bilateral language representation tended not to be consistently lateralized. Zatorre concluded that the fused rhymed task is a valid measure of cerebral language laterality. Further, magnitude of ear asymmetry appeared to be important, since large asymmetries were associated with language dominance of the contralateral hemisphere, whereas small asymmetries were more frequently associated with bilateral language.

Säisä, Silfvenius & Christianson (1990) studied visual half-field testing, but tried to increase its specificity by adjusting the stimulus presentation time to accommodate individual performance differences, as well as presenting some of the stimuli in a "mirrored" orientation. Fourteen of 16 patients were right-handed, and 13 of these had a right visual half-field advantage; 1 right-handed patient performed poorly and without any left or right half-field superiority. The 2 left-handed patients had left visual half-field superiority. These results were then compared to the Wada language findings in order to evaluate the potential for individual patient prediction, and all 13 right-handed patients with a half-field asymmetry were left cerebral language dominant. One of 2 left-handed patients was right cerebral language dominant, and the other patient displayed bilateral language, with greater right than left representation. Although the authors acknowledged the difficulty of this technique in the identification of patients with bilateral language, they were encouraged that similar non-invasive testing might eventually be substituted for the Wada language evaluation.

In summary, standard dichotic listening as well as visual half-field techniques have not proven to be sufficiently consistent to identify language dominance in individual patients. However, the studies of Zatorre (1989) and Säisä, Silfvenius & Christianson (1990) indicate that procedural modifications may ultimately make these non-invasive techniques a possible alternative to Wada testing under certain conditions.

The Crowding Hypothesis

When language functions are assumed by cerebral areas not typically associated with primarily language abilities, it is unclear what effect this language transfer may have on other cognitive functions. Teuber (1974) suggested that as the right hemisphere plays an increasingly prominent role in linguistic functions, there will be a corresponding decline in non-verbal functions typically subserved by the right hemisphere. The "crowding" hypothesis suggests a decline in cognitive abilities when "one hemisphere tries to do more than it had originally been meant to do" (Teuber, 1974, p. 73).

The Wada test has been used to classify patients to assess the effects of atypical language representation on other cognitive functions. Lansdell (1969) reported lower PIQ scores in patients with right hemisphere language dominance. Further, VIQ was less impaired with injuries prior to age 5 years. Novelly & Naugle (1985) also found lower PIQ than VIQ scores in patients with right cerebral language dominance, but this effect was present only for males. However, the males also had an earlier age of probable neurologic involvement compared to the females.

The most comprehensive studies of right hemisphere language and neuropsychological function have produced discrepant results. Mateer & Dodrill (1983) compared neuropsychological performances of left- (n=75), mixed- (n=6), and right-cerebral (n=9) language dominant patients, and observed no differences on Full Scale IQ, Verbal IQ, Performance IQ, or the Wechsler Memory Scale subtests. More "expressive" errors were reported on the Reitan-Indiana Aphasia Screening Test by the group with bilateral language.

Rausch, Boone & Ary (1991) studied left and right cerebral language dominant patients on a variety of standard neuropsychological measures. Although the right cerebral language group (n=8), all of whom had left temporal seizure onset, tended to have a more consistent history of cerebral insult prior to age 5, no difference in general level of intellectual function was present. However, this group performed more poorly than patients with left language dominance and right TLE (n=32) on the Complex Figure copy, but did not differ from left dominant patients with left TLE (n=19). This finding may be related to the smaller sample size and hence less statistical power. Verbal memory performances of left language-left TLE patients were significantly worse than the right language-left TLE group, and this was seen both pre- and post-operatively.

Strauss, Satz & Wada (1990) examined only patients with early left hemisphere injuries, and included patients with bilateral language representation (n=7) in addition to the left (n=14) and right (n=6) language dominant groups. In contrast to the Rausch et al. findings, consistently poorer performance on non-verbal tests was obtained by the patients with atypical language (i.e., mixed and right cerebral dominant). These differences were present for PIQ, Complex Figure copy, right-left differentiation, body location placement, embedded figures, mental rotation, and immediate and delayed visual reproduction memory from the Wechsler Memory Scale. No differences were present on the Benton Visual Retention Test or the delayed memory component of the Complex Figure test. In addition to poorer performance on these primarily non-verbal tests, the atypical language representation group also performed more poorly on the Boston Naming Test (confrontation naming) and Token Test (comprehension), suggesting that certain verbal skills suffer in addition to non-verbal ability.

Summary. Caution must be exercised in interpreting these reports. Due to the nature of the phenomenon under investigation, it occurs relatively infrequently, and by necessity, the reports are based upon small sample sizes. Also, the heterogeneity in the Wada test used for group assignment potentially can contribute to the discrepant results.

That small sample sizes have been relied on cannot be sufficiently stressed. It is always easier to demonstrate no group differences than a statistically significant effect. With small samples, a single, non-representative outlier will exert relatively greater effects. Further, the power of statistical tests is less with small sample sizes to study. We believe the crowding phenomenon is real, and will ultimately be verified over time as more patients are studied at different centers.

Clinical Application

In the evaluation for epilepsy surgery, assessment of language representation is an important factor frequently affecting type of procedure performed and extent of surgical resection. The determination of language representation is usually an easy task, since patients typically stop talking following left hemisphere injection, and continue to converse following right hemisphere injection. However, cases involving bilateral language may require explicit criteria for determining language representation as well as comprehensive language assessment since occasionally patients with bilateral language may be missed (e.g., Branch, Milner & Rasmussen, 1964).

The great heterogeneity in the frequency of reported bilateral language that exists from the lack of standardization across centers does not necessarily impact patient decisions. To suggest that inferences regarding language are inappropriate due to lack of standardization is analogous to stating that aphasia could not be diagnosed prior to the development of a formal battery such as the Boston Diagnostic Aphasia Examination. In most cases, the surgical approach to patients with bilateral language, most of whom have left seizure onset, is the same as left seizure patients with exclusive left hemisphere representation. The discrepancies in reported language bilaterality affect our understanding of cerebral language reorganization more, but greater consensus will undoubtedly emerge as the number of centers performing epilepsy surgery continues to grow.

In the clinical setting, we are extremely cautious in making inferences regarding other lateralized neuropsychological functions in patients with atypical language representation. For example, in 1 patient with exclusive right hemisphere language representation, a verbal memory deficit was present in conjunction with normal memory for geometric designs. Because language was restricted to the right hemisphere, one might conclude that the material-specific verbal memory deficit was suggestive of a right temporal lobe seizure onset, and the absence of a non-verbal memory impairment would suggest that standard temporal lobectomy would pose no unusual risk to recent memory functions. However, this patient eventually underwent surgery because invasive EEG recording revealed a right temporal EEG onset for his typical seizures. Following surgery, this patient has had a significant recent memory impairment, with a decline on the WMS-R General Memory Index from 84 to 58.

In contrast, we have had a patient with right cerebral language dominance but bilateral language representation, who also underwent a right temporal lobectomy. Preoperative memory testing revealed a variable pattern, with impaired selective reminding list learning and low normal digit supraspan learning. Although memory for the complex figure was normal by quantitative criteria, there was qualitative distortion on this task, which is frequently associated with right TLE (Loring, Lee & Meador, 1988). Hippocampal stimulation studies (Lee, Loring, Smith & Flanigin, 1990) revealed verbal memory decline following left hippocampal stimulation only. One-year followup neuropsychological evaluation revealed an increase in selective reminding performance as well as a decrease in Complex Figure Recall. In addition, there was a greater decline over a one half hour delay for the Visual Reproduction subtest from the WMS-R than the Logical Memory subtest.

Our personal experiences suggest that cerebral dominance for language and verbal memory are not necessarily linked. In patients with crossed aphasia (i.e., language impairment in right-handed individuals following injury to the right hemisphere), the effects of injury on visual-spatial processing have been unclear. For example, if crossed aphasia suggests a mirror cortical representation (*situs inversus*), then visual-spatial deficits associated with crossed aphasia should occur no more frequently than observed following left hemisphere damage and aphasia in the general population. However, Castro-Caldas, Confraria & Poppe (1987) reported constructional apraxia in 76% of crossed aphasics, which is much higher than the figure of 45% present in right-handed patients with left hemisphere injuries. Further, some clinical case reports of crossed aphasia have reported visual-spatial impairment (Schweiger, Wechsler & Mazziotta, 1987) while others have not (Larrabee, Kane & Rogers, 1982). Thus, given the substantial variability in visual-spatial impairment in crossed aphasia, the lack of consistency in the literature, as well as our own experience, the presence of right cerebral language representation on Wada testing significantly decreases the confidence with which inferences from neuropsychological testing are made.

2
Memory

The presence and degree of permanent memory changes following temporal lobectomy have pragmatic implications such as the ability to return to work, as well as conceptual issues involving our present understanding of memory mechanisms and the specialized role attributed to the hippocampus. The 5 original reports describing post-resection memory impairment remain the primary source of information on this potential complication, although several anecdotal reports also exist. Unfortunately, the patients' clinical histories or cognitive abilities were not always adequately described. Several papers described sub-acute memory impairment following surgery, while memory was either normal or subjectively decreased at follow-up. In other cases, mild language impairment may have been interpreted as poor memory. None of these patients displayed the magnitude of memory impairment associated with bitemporal lobectomy seen in patient H.M. Failure to distinguish memory impairment from anterograde amnesia has made it impossible to estimate the risk of post-surgical amnesia following temporal lobectomy.

The ability to predict individuals at risk for post-lobectomy amnesia remains an important goal of Wada testing. Amobarbital injection is used to model the effects of surgery on memory by creating temporary dysfunction of the temporal lobe and hippocampus while the patient is presented new material to remember. However, patients may fail Wada memory testing for reasons other than the ability of the contralateral temporal lobe to sustain memory, and methodologic limitations make the validation of this technique difficult. Consequently, performance on this test should not be considered absolute. The different options that may be pursued if a patient fails the Wada memory test are discussed.

Historical Background

Wilder Penfield at the Montreal Neurological Institute first observed significant memory impairment associated with bilateral temporal lobe dysfunction in 1951-1952. Penfield noted significant recent memory deficits following left temporal lobectomy in 2 patients who were postulated to have additional pre-existing right temporal lobe damage (Penfield & Milner, 1958). This hypothesis was confirmed in 1 patient, who at autopsy, was shown to have a pale and shrunken right hippocampus (Penfield & Mathieson, 1974).

In 1953, William B. Scoville performed bilateral temporal lobectomy for seizure control in H.M. (Scoville & Milner, 1957). In reporting H.M.'s amnesia at the Harvey Cushing Society Meeting in 1953, Scoville stated:

bilateral resection of the uncus and amygdalum alone, or in conjunction with the entire pyriform amygdaloid hippocampal complex, has resulted in no marked physiologic or behavioral changes with the exception of a very grave, recent memory loss, so severe as to prevent the patient from remembering the locations of the rooms in which he lives, the names of his close associates, or even the way to toilet and urinal... [Further] this loss was apparent in the 2 patients undergoing bilateral resection of the entire [hippocampal] complex including the hippocampal gyrus extending posteriorly for a length of 8-9 cm. from the tips of the temporal lobes." (Scoville, 1954, p. 65)

Penfield became aware of Scoville's patient H.M. in 1956:

during the preparation of our reports, William Scoville, MD, described to me the psychotic patients on whom he had operated, removing both hippocampal zones in one procedure, with the untoward results similar to my own. Our talk took place during a meeting of neurosurgeons (the proper place for discussion of unhappy results!). He had used an anterior approach to each temporal fossa and employed deep suction to remove the cerebral tissue. (Penfield & Mathieson, 1974, p. 145)

Màitland Baldwin (1956) published the first extended description of memory deficits following unilateral temporal lobectomy. Scoville & Milner's report of bilateral temporal lobectomies, which included H.M., appeared in 1957. This was followed by Earl Walker (1957), who reported marked memory impairment following temporal lobectomy in 3 patients, and Penfield & Milner's (1958) formal report of Penfield's 2 patients with significant post-surgical memory impairment. Thus, in the late 1950s the general medical and psychological community became aware of the important role mesial temporal lobe structures play in the acquisition and retention of new material, and the devastating effects of bilateral temporal lobe dysfunction. Two additional reports appeared subsequently, both suggesting risk to recent memory following unilateral temporal lobectomy (Serafetinides & Falconer, 1962; Dimsdale, Logue & Piercy, 1964).

Original Reports of Post-Surgical Amnesia

Baldwin (1956). Baldwin (1956), an epilepsy neurosurgeon from the National Institutes of Health, described 4 patients with post-operative memory deficits from his series of 65 temporal lobectomies (3L, 1R). The first case was a 40-year-old left-handed saw-mill worker with average intelligence whose seizures began at 15 years of age. Pneumoencephalography revealed dilation of both temporal horns (L>R), and skull x-rays showed a relatively small right temporal fossa. Bilaterally independent epileptiform discharges (L>R) were recorded from the temporal lobes. Following left temporal lobectomy, the patient displayed a right superior quadrantanopsia and had difficulty expressing himself. Memory for daily events was impaired. Although initially unable to identify objects, the patient later recognized objects and persons, especially if aided by cuing. He repeated himself frequently, but was oriented to place and time. Memory improved throughout the year of follow-up, and although he subjectively complained of memory difficulty 10 months after surgery, he successfully returned to work.

The second case was a 35-year-old right-handed woman of average intelligence who was a homemaker with seizures since 5 years of age. Pneumoencephalography revealed left temporal horn dilation, and skull x-rays showed a small right temporal fossa. A right temporal focus was identified with EEG. Following right temporal lobectomy, the patient displayed a left superior quadrantanopsia. She was oriented to place and time, but repeated herself frequently and had difficulty remembering daily events. Although able to remember some information with great effort, these recollections tended to be fragmented and unfamiliar. However, Baldwin reported that her memory returned to normal after 6 months.

The third patient was a 35-year-old right-handed woman who was a homemaker with a 2-year history of seizures. EEG revealed a left temporal abnormality, and a left temporal lobectomy was performed with speech and language mapping. During the 4 week post-operative period, she experienced significant memory impairment with some disorientation. Later, she returned to work as a physician's secretary, but still subjectively complained of poor memory.

The final case was a 36-year-old right-handed female homemaker with seizures since age 13. Following left temporal lobectomy, the patient complained of problems remembering events and recognizing familiar objects and persons. Baldwin reported that her memory returned to normal 4 months following resection.

Although these patients displayed memory impairment following temporal lobectomy, the decline in memory was either not permanent or was incomplete. In the first case, object identification difficulty was likely a naming rather than a memory impairment. Although poor object identification was interpreted as recognition difficulty, visual agnosia is not observed following temporal lobectomy. An upper quadrantanopsia was also present, indicating a larger resection than is now typical for language dominant hemisphere resections with increased likelihood of mild post-operative aphasia. Even if the difficulties described by Baldwin are attributed to memory, an amnestic syndrome cannot be inferred because the patient's memory improved over a 10-month period allowing his return to work.

In 2 cases, Baldwin indicated explicitly that the acute memory impairment returned to normal at later follow-up (Case 2 at 2 years; Case 4 at 4 months). In the remaining patient (Case 3), a memory deficit was reported 4 weeks post-operatively. Since this patient later served as a physician's secretary, she substantially regained the ability to remember daily activities in order to perform her job adequately. Unfortunately, longer follow-up and formal memory description was not obtained on this patient.

Scoville & Milner (1957). H.M.'s profound post-surgical amnesia firmly established the role of bilateral mesial temporal lobe structures in recent memory. H.M.'s development was normal until 7 years of age when he was knocked down by a bicycle. He sustained a laceration in the left supraorbital region and was unconscious for approximately 5 minutes (Corkin, 1984). Although the seizures are described poorly, prior to surgery H.M. averaged 10 spells per day and 1 major seizure per week. His EEG revealed diffuse slow activity with a dominant frequency of 6-8 Hz and no lateralized abnormality (Scoville & Milner, 1957). Generalized 2-3 Hz spike-and-wave discharges were present during an apparent seizure with slight EEG flattening recorded from the left central leads. On September 1, 1953, at 27 years of age, H.M. underwent bilateral mesial temporal lobe resection including prepyriform gyrus, uncus, amygdala, hippocampus, and parahippocampal gyrus. The resection extended "posteriorly for a distance of 8 cm. from the midpoint of the tips of the temporal lobes, with the temporal horns constituting the lateral edges of resection" (Scoville & Milner, 1957, p. 16).

As of 1984, he continued to have the smaller spells, but his generalized seizure frequency was reduced significantly, and he could go as long as a year between episodes (Corkin, 1984). Throughout his postoperative course, there has been a nearly complete inability to learn new information, with the exception of certain motor or procedural tasks, but intelligence and other cognitive abilities were well preserved.

In addition to H.M., Scoville & Milner (1957) described 2 other patients with significant recent memory impairment following bilateral temporal lobectomy. D.C. was a 47-year-old physician who had previously practiced medicine, but had a history of paranoia. He made a homicidal attack on his wife following the loss of a lawsuit, and was subsequently hospitalized with a diagnosis of paranoid schizophrenia. Psychosurgery was performed consisting of bilateral medial temporal lobectomy combined with orbital frontal undercutting. The posterior resection margins were approximately 5.5 cm from the tips of the temporal lobes with the inferior horns of the lateral ventricles forming the lateral edges of resection. During surgery, bilateral spiking from mesial temporal lobes was noted with some spread to the orbital frontal lobes.

Following surgery, D.C. was outwardly friendly and non-aggressive, although paranoid thought content continued. On formal testing, he obtained a Wechsler-Bellevue IQ of 122 and an MQ of 70. In describing the loss of recent memory, Scoville & Milner provided a graphic example: "At the examiner's request he drew a dog and an elephant, yet half an hour later did not even recognize them as his own drawings" (p. 17).

The third case, M.B., was a 55-year-old manic-depressive woman with normal memory who previously was employed as a clerical worker. Psychosurgery included bilateral temporal lobectomy with the posterior resections extending 8 cm from the temporal tips. Following surgery, she displayed a profound retrograde amnesia. In 1953, for example, she indicated the year to be 1950. In 1955, "her immediate recall of stories and drawings was inaccurate and fragmentary, and delayed recall was impossible for her even with prompting; when the material was presented again she failed to recognize it" (p. 17). No formal test scores were presented.

Five additional patients who underwent bilateral temporal lobe resections that extended 5-6 cm posteriorly from the temporal tip were presented. In each case, surgery was followed by moderately severe memory deficits. These patients were able to "retain some impression of new places and events, although they are unable to learn such arbitrary new associations as people's names and cannot be depended upon to carry out commissions. Subjectively, these patients complain of memory difficulty, and objectively, on formal tests, they do very poorly irrespective of the type of material to be memorized" (p.18).

These patient reports illustrate the cognitive risks associated with bilateral temporal lobectomy. However, several factors must be kept in mind when generalizing these results to current temporal lobectomy series. In the 2 cases that developed severe amnesia following surgery (H.M. and M.B.), the resections extended 8 cm posteriorly from the temporal tips. Even if performed unilaterally, this degree of resection exceeds the amount of tissue commonly removed during temporal lobectomy today--i.e., approximately 5.5 cm for language dominant temporal lobe resections and 6.5 cm if no language representation is present. The third patient developed a significant memory impairment after bilateral resection of 5.5 cm. However, this patient also received orbital frontal undercutting at the time of his bitemporal resection, and the memory deficits may have resulted from the total amount of tissue resected.

Several patients with less severe memory deficits also had other significant clinical events. One experienced a complication that resulted in coma. Another had a history of alcoholism as well as ECT that may have contributed to the memory impairment. Orbital frontal undercutting in addition to bitemporal lobectomy was performed in some patients, and the significant memory impairment may have resulted from the interaction of frontal and temporal lobe resection on memory. With the exception of H.M., these patients had significant psychiatric histories for which the psychosurgery was performed. Thus, although this report highlighted the importance of the hippocampus in memory, it cannot be concluded that hippocampal resection is necessary and sufficient for amnesia from this patient series. Careful clinical studies have documented the role of the hippocampus in recent memory in previously healthy individuals (Cummings, Tomiyasu, Read & Benson, 1984; Zola-Morgan, Squire & Amaral, 1986; Victor & Agamanolis, 1990), although other explanations for H.M.'s amnesia are possible (see Horel, 1978).

Walker (1957). Walker, an epilepsy neurosurgeon from Johns Hopkins University, described 4 cases of unilateral temporal lobectomy resulting in marked memory impairment. The first patient was a 53-year-old male with a 19 year history of seizures who underwent right temporal lobectomy. The anterior 7 cm of the temporal lobe was resected. On the fourth post-operative day, extradural and intradural clots were evacuated. Approximately 1.5 years later, after returning to work, he was found in a coma on the street and remained confused for 2 months. However, as Walker noted, "he seemed to have no memory impairment after the temporal lobectomy for over a year. It may be presumed his amnesic state was ushered in by a psychomotor attack, but whether this was associated with a vascular lesion is not known" (p. 545).

The second case was a 57-year-old female homemaker with seizures since age 46 who underwent resection of approximately 6 cm of the right temporal lobe. Following surgery, she displayed a pronounced recent memory deficit. For example, she attempted to make a purchase at a store without money. After the accompanying nurse intervened, the patient repeated the same error at a different check-out counter. The patient, unfortunately, refused all attempts at formal assessment. The memory deficit, however, was not permanent, i.e., "it seems probable that the impairment of memory has cleared to a considerable extent, if not completely, since the patient appears to be managing a large store without difficulties" (p. 546).

The third patient was a 40-year-old male who had seizures since childhood and had suffered unconscious spells and periodic confusion for approximately 12 years. The patient underwent a left temporal resection that included the anterior 6 cm of temporal lobe. Post-operatively, he had difficulty remembering daily events and the names of fellow employees. His memory was described as variable, in that "he can now remember most people and all about them at times and then the next day can't remember anything about them at all" (p. 546). Although the patient probably had borderline intelligence prior to surgery, the pattern of memory deficit was such that it was believed to be due to organic brain damage.

A final case was presented to illustrate that memory deficits seen in the initial 3 patients were due to temporal lobe lesions and not the result of diffuse, nonspecific dysfunction associated with epilepsy. A left temporal lobectomy was performed in this patient to expose an aneurysm at the carotid siphon. The temporal lobe specimen measured 4.2 cm along the inferior border, 4 cm along the superior border, and 3 cm in width. Eleven months following surgery, the patient had difficulty remembering the names of relatives and golfing partners, although she reportedly could recognize them. In addition to forgetting many details when recalling a paragraph, there was also a tendency for name substitution, such as T.S. Eliot for "T.C. Jones" and Billy Martin for "Virginia Martin." Difficulty in following a television story was also described. No other details concerning this final patient were presented.

As with Scoville & Milner's (1957) patients, it is important to consider all clinical variables when interpreting this series. Any inference regarding the relationship of the temporal lobes to memory is clouded in the first case given the presence of extradural and intradural clots following surgery. More importantly, however, this patient successfully returned to work for a year and a half prior to being found in a coma with subsequent confusion. Walker's second patient returned to managing a large store without difficulty, which suggested to Walker substantial, if not complete, return of recent memory. The third patient had borderline intelligence preoperatively, and it is difficult to know if this may have contributed to his postoperative memory deficit. Further, the memory deficit was variable, a pattern that would not be expected from a fixed neurologic lesion. In the final case in which the temporal lobe was resected to expose an aneurysm, there appeared to be a genuine deficit. However, it is not clear whether this was simply a material-specific verbal deficit (unable to remember names of friends and family but could recognize their faces) or mild aphasia, which Walker also acknowledges is a possibility. The tendency towards name substitution suggests mild aphasia.

From his series, Walker concluded that "memory disturbance occurs in not more than 10% to 15% of temporal lobe resection and may be present after removal of the lobe of either the dominant or the nondominant hemisphere" (p. 550). This is the first estimate of the prevalence of amnesia following unilateral temporal lobe resection, and these figures subsequently were used to validate results of amobarbital memory studies (e.g., Kløve, Grabow & Trites, 1969). In addition, Walker first described the relative preservation of motor skill acquisition and retention, in addition to relatively normal digit span and remote memory, an important observation for our understanding of memory mechanisms and one for which Walker usually is not credited.

Penfield & Milner (1958). Montreal's experience with significant recent memory impairment following unilateral left temporal lobectomy was published in 1958 (Penfield & Milner, 1958). The first patient presented was a 28-year-old man employed as a glove cutter with a history of recurrent seizures since age 12. His EEGs, with nasopharyngeal leads, suggested bilateral involvement that was maximal from the left side. Skull x-rays revealed a slight flattening of the left side of the vault and slight elevation of the floors of the left anterior and middle fossa. Pneumoencephalography revealed a slightly wider left lateral ventricle. Prior to surgery, he complained of being forgetful but could remember everyday events without difficulty. He obtained a Wechsler-Bellevue FSIQ of 106 (VIQ = 102, PIQ = 109), and a Wechsler MQ of 94. Surgery included resection of 5.5 cm measured from the anterior wall of the left temporal fossa along the Sylvian fissure, and 6.5 cm along the inferior temporal surface. Following surgery, the patient was aphasic and presented with a right upper quadrant field defect. Testing approximately 1 month after surgery revealed IQ decreases in Full Scale to 88 and VIQ to 80, with a less marked decrease in PIQ to 100. However, IQ assessed 6 months following surgery returned to baseline levels (FSIQ = 104; VIQ = 102, PIQ = 105). His MQ was 72, and this remained unchanged over the next 2 years.

The second patient was a 46-year-old man employed as a civil engineer. He reportedly had a single convulsion in infancy. Recurrent seizures began at age 35 accompanied by momentary lapses in his conversation. He had undergone a resection of the anterior 4 cm of the temporal lobe, sparing the hippocampus, approximately 5 years earlier. Following the first surgery, a brief period of aphasia was present. Psychological assessment prior to the second operation revealed a Wechsler-Bellevue IQ of 119 (VIQ = 125, PIQ = 110). Memory was not formally assessed. Pneumoencephalography showed moderate diffuse cerebral and cerebellar atrophy. EEG recorded prior to the first operation suggested "a well localized focal epileptogenic lesion of the left temporal lobe." EEG prior to the second surgery revealed continuous slow wave with occasional sharp waves and rhythmic 3-4 Hz activity recorded from the left temporal region. Seizure onset was lateralized to the left with an initial 6 second left-sided suppression.

In the second operation, approximately 2 cm of anterior hippocampus and the entire uncus was removed (Penfield & Mathieson, 1974). Following surgery, the patient displayed a pronounced memory impairment in which Dr. Penfield was reportedly the only hospital staff member correctly identified. In contrast to the initial operation, the second surgery caused no aphasia. One month following surgery, he obtained a FSIQ of 120 (VIQ = 129, PIQ = 107). Although the complete Wechsler Memory Scale was not administered, poor performances on Logical Memory and Paired Associate Learning were described. Five years later, the complete WMS was given, and the patient obtained a Memory Quotient of 97. The patient returned to work following surgery, but was demoted to a draftsman since he was unable to handle the administrative responsibilities associated with being a civil engineer (Milner, 1966). At autopsy, extensive atrophy of the right hippocampus was noted (Penfield & Mathieson, 1974). Approximately 22 mm of the posterior hippocampus on the resected left side was intact at autopsy.

These case reports are unique in describing patients both pre- and post-operatively. Unlike the patients described by Walker (1957), rival hypotheses to explain their memory deficits are not readily apparent. Thus, this report of 2 cases with significant memory impairment, but not dense amnesia, following surgery suggested that contralateral dysfunction may place an individual at risk for amnesia following temporal lobectomy. However, both patients returned to work following surgery, 1 as a glove cutter and the other as a draftsman. Thus, the magnitude of the memory impairment was not as severe as that of H.M., who was unable to function independently. A spectrum of memory impairment is suggested.

No other patients with marked post-operative memory impairment from the Montreal series are presented formally. However, Milner (1966) later referred to 4 additional patients who displayed memory impairment following temporal lobectomy. One right TL patient with right hemisphere language dominance developed amnesia following surgery. Three patients developed moderate memory loss following dominant hemisphere temporal lobectomy, and 1 patient developed a moderate memory deficit following left TL in which the hippocampus was spared. Formal psychological testing was not provided. Penfield & Milner (1958) concluded that although removal of the hippocampus and surrounding tissue does not typically alter memory function, significant memory impairment may be present following unilateral temporal lobectomy if contralateral mesial temporal dysfunction is present. Persistent amnesia "has been seen only in patients with electrographic or radiologic evidence of damage to the opposite temporal lobe" (Milner, 1969, p. 34).

Serafetinides & Falconer (1962). Serafetinides & Falconer (1962) studied 34 consecutive patients who had undergone right temporal lobectomy. Memory was determined by a combination of spontaneous complaints, answers to leading questions, complaints of the informant, and when available, formal tests. Seven patients (21%) displayed a recent memory impairment at follow-up, which was 2-9 years following surgery. However, 4 of these patients reportedly had persistent memory impairment pre-operatively. Six of 7 patients had post-operative epileptiform activity recorded using sphenoidal electrodes from the left, non-operated, temporal lobe. The authors interpreted their findings to be consistent with Penfield & Milner (1958), in that memory impairment was present in patients with contralateral mesial temporal lobe dysfunction.

Although Serafetinides & Falconer's patients presented with clinical memory impairment, they did not develop an amnestic syndrome. Four of the 7 cases with post-operative memory impairment had impaired memory prior to surgery. In 4 of 6 patients who were administered the Wechsler Memory Scale both pre- and post-operatively, a decline was noted in only a single patient, and in this patient, "his auditory learning ability was intact." Unfortunately, performance levels on the WMS were not presented. The authors concluded that "the type of memory defect we are now considering does not correlate with the more formal psychological test results" (p. 254). The authors further stated that the memory deficit was "usually compensated for," illustrating an absence of an amnestic syndrome.

Dimsdale, Logue & Piercy (1964). Persistent amnesia following right temporal lobectomy in a 53-year-old woman with no evidence of contralateral dysfunction was reported by Dimsdale, Logue & Piercy (1964). The patient experienced her first seizure at age 25. Grand mal attacks occurred several times per year, but increased in frequency to several times per month. Minor seizures without loss of consciousness initially occurred rarely, but progressed to several per day. At age 37, she was hospitalized for depression. Her husband noted that she often became depressed during the winter months "and had for years been a difficult woman to manage, being highly sensitive to criticism, quarrelsome, especially with neighbors, and had alienated most of her friends." Her WAIS IQ prior to surgery was 100, (VIQ = 101, PIQ = 99) and MQ was 80.

Her right temporal lobectomy extended 6 cm from the tip. Following surgery, no change in consciousness or motor activity was noted. She became confused during the week following surgery and was admitted to a mental hospital due to paranoia and amnesia. Three weeks post-operatively she obtained a pro-rated Full Scale IQ of 102, and an MQ of 101. However, the authors discussed significant difficulty with material retention and slow learning with no savings on relearning. Similar impairments were present for visual design learning. She was reportedly unable to remember the previous day or recognize individuals with whom she had had previous contact.

Some curious findings were also present. For example, she appeared to know the date to within a few days and was reportedly oriented for time and place 8 months post-operatively. This level of temporal orientation would not be expected in a patient unable to learn new information. Further, her Memory Quotient actually increased 21 points to the average range. An increase of this magnitude suggests a significant functional component to her memory test scores. Although the MQ may not be as sensitive to memory deficits as examining delayed recall, the MQ has been employed routinely to document the memory deficit associated with Korsakoff's disease (e.g., Victor, Herman & White, 1959).

An interesting verbal/visual memory dissociation was also present. For example, she could recognize the name of one author, but could not recognize him in person. "I recognize your name but I don't know your face. Would you be his brother?" This type of nearly correct response, seen also with her temporal orientation, raises further concern regarding the etiology of her post-operative memory deficit. Since there was no post-mortem examination, the case was tested prior to CT and MRI, and a significant psychiatric history exists, it is difficult to attribute the post surgical memory deficit to a single cause.

Conclusions. Although significant memory impairments may occasionally be produced by unilateral temporal lobe resection, review of the original literature suggests that when a noticeable memory deficit is present, it is not as extreme as the amnesia observed in patient H.M. Memory impairment was described during the immediate post-surgical recovery, but often, memory returned to normal or was only subjectively decreased with longer follow-up. In some instances, mild aphasia may have been considered memory impairment. In other cases, patients returned to work at their previous occupation. The cases from Montreal indicate that contralateral dysfunction places individuals at risk for significant memory impairment following temporal lobectomy. However, the description of these cases suggests a spectrum of memory impairment rather than amnesia.

We believe the memory impairment seen occasionally following temporal lobectomy became synonymous with amnesia due to the association with patient H.M. (e.g., Milner, 1966; Jones-Gotman, 1987). H.M. typically is described to illustrate the severe and permanent amnesia associated with bilateral hippocampal dysfunction. Patients with memory impairment following unilateral resection and presumed contralateral dysfunction are also commonly cited to illustrate the importance of the hippocampus in memory function. Since hippocampal involvement has been postulated as the mechanism for both types of memory impairment, and since the cases of memory impairment following unilateral temporal lobectomy and the amnesia displayed by H.M. typically are presented conjointly, there has been a tendency to treat the deficits themselves as functionally equivalent.

The likelihood of amnesia following unilateral temporal lobectomy cannot be determined from the original studies. Nevertheless, Penfield & Milner's patients document the risk for post-resection memory impairment if contralateral temporal lobe dysfunction is also present, and patients with atypical language representation may invalidate the interpretation of standard neuropsychological evaluation. Less severe impairment in recent or material-specific memory may remain a significant consideration in individuals who rely heavily on memory for their occupational success. Consequently, Wada memory testing is routinely employed prior to anterior temporal lobectomy.

Rationale for Wada Memory Assessment

When Penfield & Milner (1958) postulated that an occult contralateral temporal lobe lesion contributed to the marked memory deficit occasionally seen following unilateral temporal lobectomy, it became necessary to develop techniques capable of predicting hippocampal damage contralateral to proposed surgery. Milner, Branch & Rasmussen (1962) first included memory testing during amobarbital anesthesia to predict the presence of contralateral mesial temporal lobe dysfunction. Their rationale was that by producing a state of temporary, reversible dysfunction ipsilateral to the side of proposed surgery, the potential effects of temporal lobectomy on memory function could be modeled prior to surgical resection.

Three tacit assumptions are associated with intracarotid amobarbital memory testing: 1) pharmacologic inactivation of a single temporal lobe does not by itself create a global memory defect, 2) critical memory regions resected during temporal lobectomy are functionally inactivated by internal carotid injection, and 3) if mesial temporal lobe structures contralateral to amobarbital injection are sufficiently dysfunctional at baseline such that resection of the epileptogenic temporal lobe would create an amnestic syndrome, then injection will produce a transient amnestic state due to temporary bilateral mesial temporal lobe dysfunction. If patients fail to recall/recognize a sufficient number of stimuli presented following the injection ipsilateral to seizure onset, they are believed to be at risk for a post-resection amnesia. Three early reports established the feasibility of this procedure.

Milner, Branch & Rasmussen (1962). Milner, Branch & Rasmussen (1962) showed patients drawings of objects to be remembered both before and approximately 3 1/2 minutes after injection of 200 mg sodium amytal (10% solution). Pre-injection item recall was assessed immediately prior to presentation of the 2 post-injection pictures. Post-injection recall was tested after a non-specified distraction. If unable to spontaneously recall either pre- or post-injection objects or to indicate their use, recognition memory was tested employing 5 foils. The presence of any error, presumably either omission or false positive response, was considered memory failure for that injection.

A series of 50 consecutively tested patients was reported. However, not all patients were studied bilaterally; 44 received language dominant hemisphere injections and 46 non-dominant hemisphere injections. Memory impairment was present after 5 (11%) dominant and 7 (15%) non-dominant hemisphere injections. Memory failure occurred after injection contralateral to the epileptogenic lesion in 11 cases. In the final case, the amobarbital reportedly was injected "rapidly," creating bilateral neurologic dysfunction, and the results were not considered valid.

One patient with bilaterally independent seizure onset was used to illustrate the clinical utility of this technique. Left hemisphere injection caused a clear aphasia, yet did not affect memory. In contrast, right hemisphere injection did not alter language function but produced "an unmistakable anterograde amnesia." Based on these results, a standard left temporal lobectomy including hippocampus and hippocampal gyrus was performed and no post-operative memory loss was noted.

Milner discussed 2 potential technical limitations of Wada memory assessment during this procedure. First, the region of interest may not be perfused reliably with carotid artery injection and thus, hippocampal structures may not be anesthetized. The second limiting factor was the restricted time available for assessment. Milner presented these data as preliminary, however, and realized that refinement of the technique would be necessary.

Kløve, Grabow & Trites (1969) employed both pre- and post-injection item presentation to test memory. Of 20 patients evaluated with a 180 mg injection ipsilateral to seizure onset, 4 patients failed the memory test. As Milner (1969) observed when discussing this paper, however, no criterion of memory failure was employed, and only subjective consensus served to classify patient performances. None of the patients in their series undergoing temporal lobectomy experienced a persistent memory deficit as determined by neuropsychological assessment, and the 20% failure rate was contrasted with an *expected* incidence of post-resection memory deficits of 10-20% (e.g., Walker, 1957). Unfortunately, the expected frequency of post-resection memory impairment used to validate their findings was based upon subjective experience rather than performance of patients following temporal lobectomy. These data were presented only in abstract form, and no details regarding mean performance levels were described.

Blume, Grabow, Darley & Aronson (1973) presented 9 pictures following amobarbital injection of 125 mg. Patients were tested for their ability to recall and recognize the material after the medication had worn off. Failure was defined as recall of fewer than two-thirds of the items. Unfortunately, only 1 patient received bilateral injections. None of the patients failed the test, and no post operative amnesia was described in any of the patients following temporal lobectomy. Because no patient in the above reports who successfully performed the amobarbital memory test displayed clinically apparent memory deficit after temporal lobectomy, this technique has been adopted widely to screen patients with possible bilateral hippocampal dysfunction prior to unilateral temporal lobectomy (Rausch, 1987).

General Assessment Strategies

Two principal approaches to test memory during amobarbital assessment have evolved. One method employs an ongoing stimulus-distractor-recognition format in which a stimulus is presented and the subject asked to identify it. Following a distractor, recall of the original stimulus is obtained. The time until successful return to baseline memory levels may be used as the index of memory performance, or a memory test may be administered after the medication effects have worn off.

The other technique involves discrete item presentation during the period of hemispheric anesthesia, and then testing recall after amobarbital anesthetization has worn off. This approach was developed in Montreal and subsequently employed by Kløve, Grabow & Trites (1969) and Blume, Grabow, Darley & Aronson (1973). Items are presented at certain times throughout the procedure. After medication effects have receded, memory is tested. Free-recall failure is commonly not considered memory failure, and evidence exists that recognition rather than recall is a more reliable measure of hemispheric memory (Christianson, Säisä & Silfvenius, 1990; Walker & Laxer, 1989). Snyder & Novelly (1991) reported that of the 55 epilepsy surgery centers surveyed, 19 (29%) employed a stimulus-distractor approach, whereas 39 centers (71%) employed discrete item presentation.

Few systematic studies examining Wada memory performance have been conducted. In her description of amobarbital memory testing and its role in identifying patients at risk for post-surgical global memory loss, Rausch (1987) referred to 2 proceedings of scientific presentations (Kløve, Grabow & Trites, 1969; Milner, Branch & Rasmussen, 1962), 2 book chapters (Engel, Crandall & Rausch, 1983; Milner, 1975), and 1 peer-reviewed journal article (Rausch, Fedio, Ary, Engel & Crandall, 1984). The book chapters were from the same groups that had published 2 of the other 3 reports on memory testing. Thus, references from only 3 different epilepsy surgery centers were cited. In other reports of Wada memory testing, the sample sizes of unilateral TLE patients have been small (e.g., n=12, Fedio & Weinberg, 1971; n=9, Silfvenius & Blom, 1984). Thus, inclusion of the Wada memory component has been relied upon despite limited validation from independent surgery centers, and only recently have more tightly controlled reports appeared in the literature.

Memory Studies Involving Continuous Material Recognition

Fedio & Weinberg (1971) tested memory by presenting photographs of familiar objects alternating with the word "AND." The task was to identify "AND," to name each object, and then to recall the name of the immediately preceding object. Thus, when presented with each new object, the subject was to name it and then recall the name of the previously displayed object that appeared prior to "AND."

Twelve left cerebral language dominant patients were studied with 125 mg dosage. Seizure onset was lateralized to the left hemisphere in 6 patients and to the right in 6. Following left carotid injection, the first correct recall was observed at approximately 3 minutes, although memory remained impaired relative to baseline until 8-10 minutes after injection. Following right hemisphere injection, memory was impaired but for a shorter duration. A significant left vs. right injection effect was present for object recall (p < .001), with poorer performance observed following the left hemisphere studies. Eight patients could not recall their aphasia, and all patients reported that the left hemisphere injection created a more severe memory deficit. Of particular importance from a clinical perspective was the absence of differential memory performance as a function of seizure onset laterality. Fedio & Weinberg stated that their patients did not have abnormal EEG discharges from the hippocampal region; however, depth electrodes were not used. Since there was no effect on memory for side of injection relative to seizure onset (e.g., absence of contralateral/ipsilateral effect), their memory task appears insensitive to bitemporal dysfunction.

Engel, Rausch, Lieb, Kuhl & Crandall (1981) administered 125 mg injections to 7 temporal lobectomy candidates and employed a continuous performance/recognition task to assess memory. Memory was based upon the recognition of material presented 3 to 4 minutes post-injection, and assessed approximately 12 minutes following the injection. Three of 7 patients displayed poor recognition memory contralateral to the side of eventual resection (2 R,

1 L), with adequate memory performance following ipsilateral injection. Of the remaining 4 patients, 1 received a single injection ipsilateral to surgery and demonstrated adequate memory while the other 3 patients displayed adequate memory following both injections. The criteria for passing or failing were not included, and their treatment of false positive responses, if any, was not described.

Rausch, Fedio, Ary, Engel & Crandall (1984) examined behavioral recovery time following unilateral amobarbital injection of 125 mg in 17 TLE patients (6 L, 11 R). Common objects, written words, and unfamiliar forms were presented in a continuous recognition format. Left vs. right hemisphere injection effects were different for left TLE and right TLE patient groups. This differential recovery time as a function of seizure laterality was present for short-term recognition of pictures, words, and forms. However, when examining memory following return to baseline, an ipsilateral vs. contralateral effect was present only for picture recall. Word recognition was poorer following left hemisphere injection, but did not differ between the 2 TLE groups. No effect for form recognition was detected.

It is worth examining further the statistical interaction for picture recall. According to her figures, presented in Figure 2-1, there was little performance asymmetry for injection side in the left TLE patients. In contrast, patients with right TLE performed more poorly following left (contralateral) hemisphere injection. Thus, the statistically significant interaction was due to the ipsilateral vs. contralateral performance asymmetry for right TLE patients only. Consequently, this approach appears best suited to assess only the functional memory capacity of the left temporal lobe in right TLE patients.

Conclusions. The continuous performance/recognition memory paradigm has critical limitations for identifying risk for post-surgical amnesia. Patients may be unable to perform secondary to transient aphasia associated with language dominant hemisphere injection. If unable to execute specific tasks due to a comprehension deficit, it cannot be known if the stimuli or response arrays are being adequately attended. Similarly, acquisition and retrieval are confounded since both are being assessed during the period of hemispheric anesthesia. Unless recognition is assessed following anesthesia, this approach does not evaluate the retention of material over time.

Even when employing post-drug recognition scores, several problems remain. The only differential effect as a function of ipsilateral vs. contralateral injection reported by Rausch, Fedio, Ary, Engel & Crandall (1984) was seen only in right TLE patients. However, non-dominant temporal lobectomies generally are felt to produce a much smaller risk to recent memory than when resection of the language dominant temporal lobe is performed (Milner, 1966).

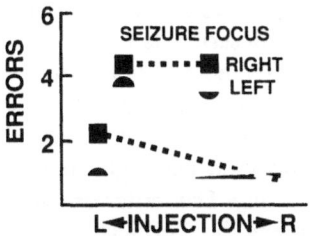

Figure 2-1   Mean number of errors for recognition of common object pictures.   Modified from Rausch, Fedio, Ary, Engell & Crandall, 1984; Modified with permission, Little, Brown & Co.

Memory Studies Involving Discrete Item Presentation

The Montreal procedure underwent slight evolutionary change over the years. Initially, 2 pre-injection and 2 post-injection line drawings were the stimuli to be remembered. A pre-injection sentence and a post-injection nursery rhyme were later added (Milner, 1972). Approximately 3 minutes post-injection, the patient was asked to recall the pre-injection material. Although some aphasia was frequently present, creating difficulty in recall of the pre-injection material, Milner (1972) asserted that "the critical test, however, is for anterograde amnesia" (p. 441). To test for anterograde amnesia, 2 line drawings and a nursery rhyme were presented during hemispheric anesthesia, and the patient was instructed to name the pictures, repeat the rhyme, and remember the items. Free recall was assessed after a distraction period and, if necessary, picture recognition was obtained. At the time of memory testing, motor strength was generally normal although its complete recovery was not required.

Anterograde amnesia for the post-injection material was observed only 27 times following 226 injections (Milner, 1972). Amnesia occurred in 25/110 injections contralateral to the lesion, but in only 2/116 ipsilateral injections. When cases with only 1 error were eliminated (equivocal memory failure), 18 patients failed the amobarbital memory test, all during injection contralateral to the seizure focus. In 1975, Milner recommended that at least 2 recognition errors be used as evidence of memory impairment, stating that "anterograde amnesia occurred 18 times, always after injection contralateral to a known temporal lobe lesion" (p. 317).

In 1987, the amount of amobarbital administered had decreased from 200 mg to 175 mg (Jones-Gotman, 1987). Two additional items (real object, printed word) were presented during the drug effect to supplement the line drawings and nursery rhyme. For example, the patient was shown a can-opener when he/she had successfully named 2 objects during language testing. The patient was instructed to name its color and to hold it. The other item was a printed word of a concrete noun (e.g., coffee pot). The subject was asked to read the word and describe its use.

Memory for material presented during the drug effect was tested after all medication effects had receded. Recognition memory of the object was obtained employing a foil. Free recall for the other 4 items was tested, followed by a yes-no recognition paradigm and, when necessary, a forced choice among multiple items. The criterion of failure was 2 errors on recognition assessment among the 5 critical items.

Using this procedure, patients with a clear unilateral seizure onset had a 41% failure rate following contralateral injection, but only a 15% failure rate following ipsilateral injection. In contrast, patients with bilateral seizure onset had a 60% failure rate contralateral to proposed surgery and a 50% failure rate on the side of the proposed resection. The number of subjects studied was not identified.

Silfvenius and colleagues (Silfvenius, Blom, Nilsson & Christianson, 1984; Silfvenius & Blom, 1984) examined Wada memory following injection of 150-170 mg amobarbital in 18 patients. Patients included those with unilateral and bilateral temporal lobe seizures, and others with frontal lobe seizures. Prior to injection, patients were presented 4 types of material-- a color, a numeral, a simple sentence, and tactually presented common objects. Following disappearance of contralateral EEG slowing, typically 30-50 seconds post-injection, 6 words and 6 pictures were presented. Additional objects were placed in the ipsilateral hand, and the patient was asked to identify them without looking.

Pre-injection item recall was obtained both during drug effect and after return to baseline. However, not all patients received all tests. During drug effect of the language dominant hemisphere, the pre-injection color was recalled during only 4/15 dominant hemisphere injections, number recalled in 4/15 cases, and sentence recalled in 0/12 cases. Following recovery, performance improved with 14/15 patients recalling the color, 12/15 patients recalling the number, and 10/12 patients recalling the sentence. Thus, sentence recall appeared most affected by the presence of linguistic disturbance.

Following injection to the non-dominant side, the color was recalled following 12/14 injections, number recalled in 12/14 patients, and sentence recalled in 10/12 cases. After the medication effects had receded, all 15 patients recalled the color following non-dominant hemisphere injection, 14/15 patients recalled the number, and 11/12 patients recalled the sentence (Silfvenius & Blom, 1984).

Recall of post-injection words following dominant hemisphere injection averaged 0.9 words; following non-dominant injection, recall averaged 3.2 words (Silfvenius, Blom, Nilsson & Christianson, 1984). Similarly, picture recall averaged 2.1 following dominant hemisphere injection and 3.5 pictures following non-dominant hemisphere injection. However, since patients with non-temporal foci were included, it is important to examine patients with temporal lobe seizures. Of the 3 left TLE patients, all were unable to recognize any of the 6 words presented during dominant hemisphere anesthetization, although 1 patient recognized 4/6 pictures and 2 others recognized a single picture. In contrast, 2 of these patients recognized at least 2/6 words and 2/6 pictures following non-dominant hemisphere injection. The final left TLE patient recognized no pictures, although this was the patient who demonstrated adequate picture recall (4/6) on the contralateral study. This pattern is opposite that which would be predicted from a left seizure focus since right hemisphere injections should have created greater memory impairment.

The 5 right TLE patients displayed performance patterns similar to left TLE patients, with superior performances following non-dominant hemisphere injection. Two patients showed at least partial recognition of the pictures with scores of at least 3 after non-dominant injection, 1 of which also recognized a single word. The remaining 3 right TLE patients were unable to recognize any of the material. Following dominant hemisphere injection, all 5 patients recognized at least 2 pictures, and 3 patients recognized at least 3 pictures. The 2 patients with poorer picture recognition were unable to recognize any words, although the other 3 patients recognized at least 3 of the presented words. Overall, the expected performance impairment seen with contralateral injection was observed in the right temporal patients, with memory impairment seen following left hemisphere injection.

Verbal recall of tactually presented material prior to injection was impaired due to aphasia following dominant hemisphere injection. New material was presented under drug influence when adequate finger movements demonstrated some return of function. Following recovery, 3/12 patients could verbally identify the material. In contrast, 6/8 patients could recall the pre-injection items during the period of non-dominant hemisphere anesthesia. Recall appeared greater following nondominant hemisphere injection. The presence of aphasia appeared to be affecting the results, with the expected performance asymmetries present for only the right TLE patients. As with studies employing continuous memory recognition, Silfvenius demonstrated a convincing performance asymmetry in the expected direction only for right temporal lobe patients (Silfvenius & Blom, 1984).

Lesser, Dinner, Lüders & Morris (1986) examined memory performance for discrete objects presented during the mute state commonly seen immediately following left hemisphere injection. Patients received injections of 165 mg amobarbital. Following left hemisphere injection, a variety of behavioral effects were observed with some patients closing their eyes, attempting to get off the angiography table, or becoming agitated, whereas others would lie quietly with eyes open or closed. Immediately after contralateral hemiplegia, the patient's eyes were held open when necessary and, when possible, objects were placed in the patient's ipsilateral hand. Objects were presented throughout the initial confusional period, and due to individual differences in responsiveness, the number of items presented varied.

Object recognition was tested by sequential recognition after recovery from dominant hemisphere injection. Individual items were interspersed with foils, although it was not stated how false positive recognitions were treated in the overall scoring strategy. These authors found that 18/24 patients with left temporal seizure onset and 4/12 with right temporal seizure onset could correctly recognize at least two-thirds of the objects presented. Further, 17/20 patients who could recognize all objects had left seizure onset, and 6/6 patients who failed to recognize any objects had right temporal lobe seizures. Lesser, Dinner, Lüders & Morris (1986) concluded that consciousness can be retained despite transient disruption of left hemisphere language mechanisms (see Attention Chapter). In addition, using discrete items in TLE patients following left hemisphere injections, the authors convincingly demonstrated that a differential memory effect occurs as a function of seizure laterality. Because their goal primarily was designed to demonstrate recall of material during the early mute/confusional state, and because of the absence of a similar state following right hemisphere injection, early objects were not presented during injection of the right internal carotid artery.

Powell, Polkey & Canavan (1987) employed discrete item presentation to assess memory in 27 temporal lobectomy candidates (Left TLE = 14, Right TLE = 13). Their technique of drug administration differed from most centers, with amobarbital injected at a rate of 25 mg every 5 seconds until the development of contralateral hemiplegia. Before drug administration, patients were presented with a simple sentence (e.g., "Tom's dog ran down the road with a bone in its mouth") and asked to repeat it 3 times. Following development of hemiplegia, patients were asked their name and address, a procedure that the authors used to test both language and memory functions. Three to 5 line-drawings were administered to assess naming. Three to 5 printed words were presented for reading, and 3-8 household objects for memorization. After recovery from drug effect, free recall was requested first. This was followed by recognition memory, which was assessed with target items interspersed with foils. Finally, the patient was asked for free recall of the pre-injection sentence.

The clinical interpretation was based subjectively on the relative performances of the 2 hemispheres, and no absolute criteria were imposed to account for patient variability. Several clinical patterns were described. For example, the authors suggested adequate recognition ipsilateral to seizure onset might be 5 correct recognitions (total = 9-18) with no false positives, vs. 1 correct recognition and 1 false positive seen on the contralateral side. If 3 or fewer correct recognitions were obtained from each hemisphere, the patient was generally considered at risk for post-resection memory loss with neither hemisphere able to sustain global memory. In other patients, adequate memory was observed following both injections, although a slight asymmetry of function may have been noted. False positive responses during recognition cast doubt on the performance validity, and their presence invalidated the entire memory performance for that injection. The patient was considered to have "failed" the memory component. Powell, Polkey & Canavan interpreted poor amobarbital memory performance as a strict contraindication to temporal lobectomy. In the series of 27 patients, 10 patients (37%) were not operated upon because of their amobarbital memory results (Left TLE = 5, Right TLE = 5).

The 3 patients who displayed bilaterally equal and adequate memory merit special consideration. When impaired memory is observed following injection contralateral to seizure onset, and normal memory is present following ipsilateral injection, it is assumed that surgery presents no threat to memory. However, if memory is represented bilaterally, as evidenced by good memory performance following each injection, it is possible that resection may produce a significant memory loss because functional tissue may be included in the resection. However, Powell, Polkey & Canavan (1987) reported 3 patients in whom both hemispheres appeared to be participating in the formation of new memories and left temporal lobectomy was performed. Two patients increased their memory function, and 1 patient experienced "some loss" of memory, i.e., Wechsler MQ decreased from 74 to 63. Patients in this series were tested 1-6 months following surgery. Engel, Rausch, Leib, Kuhl & Crandall (1980) similarly described 3 patients with bilaterally intact amobarbital memory results who underwent right temporal lobectomy without significant alteration of recent memory function. Thus, the presence of adequate memory bilaterally should not be interpreted as necessarily indicating a significant risk for post-operative memory deficit.

Several difficulties are present, however, when interpreting the Powell, Polkey & Canavan results. No formal criteria were employed to evaluate memory performance. Further, asking patients their name and address to test memory assesses overlearned responses from remote memory rather than the critical ability to learn new information. Despite these limitations, Powell, Polkey & Canavan (1987) directly addressed the issue of false positive responses during recognition memory assessment. Their approach was to consider memory performance completely invalid in the presence of false positive responses. Since false positive response frequency within subjects was not presented, it is difficult to know to what degree their decision contributed to the rate of memory failure ipsilateral to seizure onset, thereby precluding surgery. However, false positives cannot be ignored since a patient may not only correctly recognize the target material, but also recognize items never presented. Loring, Meador & Lee (1989) reported that false positive errors occurred more frequently following left hemisphere injection and an inverse relationship was presented between false positive response and baseline verbal memory. False positive responses were interpreted as reflecting memory impairment rather than failure to suppress incorrect responses secondary to poor self-monitoring skills.

Rausch, Babb, Engel & Crandall (1989) described memory performance using discrete items in 30 patients with unilateral temporal lobe seizure onset (Left TLE = 13, Right TLE = 17). Following administration of 125 mg amobarbital, discrete memory items were presented during the period of drug-induced unilateral EEG slowing and hemiparesis. Only functionally appropriate stimuli were presented, i.e., only materials that could be easily processed by the hemisphere being evaluated. For example, "three pictures of common objects and a geometric shape that could be encoded by multiple modalities were used to assess the functional integrity of either hemisphere. However, verbal items were not considered critical for assessing the nondominant hemisphere after injection of the language-dominant hemisphere" (p. 784). The total number of items presented was not mentioned in their report, nor was a detailed description of the pool of functionally appropriate items for the right hemisphere included.

Memory was assessed by free recall followed by multiple choice recognition if necessary. Although foils were included in the recognition portion, control for guessing and false positive responses was not described. A passing score was correct recognition of at least 67% of the critical items. Nineteen of 30 patients failed memory performance following injection to the hemisphere contralateral to the seizure onset. In contrast, all patients recognized at least two-thirds of the material following ipsilateral injection. These results confirmed the findings of Lesser, Dinner, Lüders & Morris (1986) demonstrating that left TLE patients can perform recognition assessment following left hemisphere injection.

Of particular interest was the finding that when significant cell loss (>80%) was observed in the mesial temporal lobe structures ultimately resected, 5/6 patients failed to recognize at least 67% of the presented material following contralateral injection. In contrast, memory performance following contralateral injection in patients with less than 80% neuron loss was more variable, with only 14/24 patients failing the memory component. Further, several patients with relatively little cell loss failed the memory component, whereas some patients with greater neuronal loss performed satisfactorily. Rausch, Babb, Engel & Crandall (1989) suggested that performance decrements contralateral to the proposed surgery demonstrated the presence of function that could sustain memory following temporal lobectomy. Thus, contralateral injection provides indirect evidence for absence of risk to recent memory following temporal lobectomy.

Some relationship between cell densities and Wada memory performance was demonstrated by Sass, Lencz, Westerveld, Novelly, Spencer & Kim (1991). Patients who failed to demonstrate adequate memory following injection contralateral to the seizure focus were shown to have relatively fewer cells in the CA3 hippocampal subfield than patients with adequate memory. Further, a significant correlation was present between Wada memory performance of the contralateral side and CA3 cell density. As with the Rausch, Babb, Engel & Crandall (1989) report, there was a less strong relationship in patients with moderate neuron loss. Sass et al. suggest that specificity exists in this hippocampal subfield given the relationship between Wada memory performance contralateral to the seizure focus and CA3 cell loss.

Table 2-1 Memory performance of patients with temporal lobe seizures

	Words (n=6)	Pictures (n=6)
Left TLE		
L Injection	1.26 (21%)	1.86 (31%)
R Injection	2.46 (47%)	2.46 (47%)
Right TLE		
L Injection	0.30 (5%)	1.86 (31%)
R Injection	2.82 (47%)	3.24 (54%)

Adapted from Aasly & Silfvenius (1990)

The confounding effects of language impairment on memory tests with a prominent verbal component has been illustrated in several reports from the Umeå program in Sweden. Aasly and Silfvenius (1990) examined memory for single printed common nouns, as well as memory for pictures presented during "early" and "late" portions of the study. The authors used chi-square analysis rather than a parametric approach to data analysis, making it difficult to determine whether any statistical interactions for lesion laterality as a function of hemisphere injected were present. Nevertheless, across the entire series of patients, which included non-temporal as well as temporal lobe seizure patients, free recall was superior following injection to the non-dominant hemisphere. Further, even when examining recognition memory effects, which minimize the effects of language impairment on memory performance following left hemisphere injection, a consistent difference between the language dominant and the non-dominant hemisphere was observed. Higher performance was seen following non-dominant hemisphere injection. This pattern could not be attributable to the inclusion of non-temporal lobe patients in the analysis. As can be seen in Table 2-1, even when restricting the sample to TLE patients, there is consistently higher performance following right hemisphere injection independent of seizure onset. This finding is consistent with their earlier report (i.e., Silfvenius & Blom, 1984). The only statistically significant difference they report in the temporal lobe patients was for word recognition, with poorer performance present following left hemisphere injection in right TLE patients. Aasly and Silfvenius acknowledged that the timing of their early items was "very early in comparison to those reported in other studies." However, their comparison of early vs. late item memory performance revealed comparable levels, suggesting the material could be presented throughout the procedure, and even following recovery of some motor and speech functions.

In a different report from the same center, Christianson, Säisä & Silfvenius (1990) investigated material-specific memory by employing concrete and

Table 2-2 Memory performance for different types of stimuli (n=3 for each condition) for TLE patients

	L-TLE	R-TLE
Concrete Words		
L Injection	0.21 (7%)	0.24 (8%)
R Injection	0.66 (22%)	0.75 (25%)
Abstract Words		
L Injection	0.00 (0%)	0.45 (15%)
R Injection	0.12 (4%)	0.45 (15%)
Common Objects		
L Injection	1.89 (63%)	0.75 (25%)
R Injection	0.60 (20%)	1.14 (38%)
Geometric Figures		
L Injection	0.90 (30%)	0.39 (13%)
R Injection	0.45 (15%)	0.63 (21%)
Faces		
L Injection	0.90 (30%)	0.51 (17%)
R Injection	0.12 (4%)	0.39 (13%)

(Adapted from Christianson, Säisä & Silfvenius, 1990; with permission)

abstract words, common object pictures, geometric figures, and faces as stimuli. The statistical analyses involved both left and right TLE patients, as well as a group of patients with left frontal seizure onset, making direct comparison of the 2 TLE groups difficult. However, by examining performance patterns of the 2 temporal lobe groups directly (see Table 2-2), it is clear that neither concrete nor abstract words selectively differentiated left and right TLE groups as a function of injection laterality, with poorer word memory generally observed following left hemisphere injection. In contrast, memory for common object pictures revealed the expected double-dissociation, with poorer memory obtained following injection contralateral to the seizure focus in both left and right TLE groups. A similar, but smaller effect was present for geometric figure memory. Examination of Table 2-2 reveals that the only stimulus material with ipsilateral injection recall greater than 1 (maximum = 3) was for common object picture. Thus, for making clinical decisions, object pictures appeared to be the best measure of hemispheric memory function. This confirms their earlier observation made on fewer subjects, in which picture recognition displayed the expected double dissociation as a function of seizure onset laterality and side of amobarbital injection.

We studied both early and late memory components of the Wada evaluation in 57 TLE patients employing incremental injections of 75-250 mg (average = 129; Loring, Lee, Meador, Flanigin, Smith, Figueroa & Martin, 1990). Immediately following demonstration of hemiplegia and evaluation of eye movements, the patient was requested to execute a simple command (e.g., "touch your nose"). Eight common objects (i.e., "early" items) were then presented for approximately 4 seconds each, and the object names repeated twice to the patient. The patients' eyes were held open as necessary, and the objects were presented in the visual field ipsilateral to the hemisphere injected. After return of minimal language function, "late" items were presented. The late items included naming 2 objects. If the patient was unable to name an item, its name was repeated twice to the patient. A nursery rhyme was read to the patient, and the patient was requested to repeat it. Two visual discrimination items also were included to provide balance against the primarily verbal aspects of the other late memory stimuli.

Following return to baseline levels (e.g., motor strength, language comprehension and repetition), recognition memory for the early objects was tested with 16 randomly interspersed foils. If unable to recall freely the 2 late objects, the names of 5 objects were read to the patient for recognition assessment. Similarly, if unable freely to recall the rhyme, recognition for the rhyme was assessed. Only recognition assessment was analyzed, and free recall was scored as evidence of correct recognition. A multiple choice recognition assessment was also obtained for the 2 designs. Both early and late item memory components were sensitive to bilateral temporal dysfunction by virtue of significantly poorer overall group performance on the injection contralateral to seizure onset.

Despite the statistical sensitivity to bilateral temporal lobe dysfunction, the individual predictive ability of Wada memory results was poor. The authors employed relatively conservative criteria for establishing the performance threshold for memory failure. Using their criteria, 13 patients failed either the early or late item recognition memory tests. However, based upon performance on other tasks, i.e., electrical hippocampal stimulation with simultaneous memory testing and memory assessment following hippocampal cooling at the time of surgery, 10 patients received standard temporal lobectomies including hippocampus. None of the 10 patients developed an amnestic syndrome.

Comment. The timing of stimulus presentation is potentially an important variable. When discrete rather than continuous item presentation is employed, a tradeoff between Type I and Type II statistical errors may be present. A Type I error occurs in statistical hypothesis testing when the null hypothesis is rejected incorrectly and group differences are inferred when, in fact, chance fluctuation alone is responsible for the observed effect. In contrast, a Type II error occurs when it is incorrectly concluded that no group difference exists

Table 2-3 Full length reports describing memory performance following amobarbital administration

Study/Sample/Dosage	Stimuli	Memory Task	Memory Results	Comment
CONTINUOUS RECOGNITION				
Fedio & Weinberg (1971) 6 L-TLE 6 R-TLE 125 mg	Photographs, word "AND"	Identify "AND," name object, recall name of immediately preceding object	Left hemisphere/right hemisphere differences present for naming and and recall	No memory effect seen as function of seizure laterality
Engel et al. (1981) 2 L-TLE 5 R-TLE 125 mg	Naming, Reading, and Design Matching	Stimuli matched, distractor, then recognition. Post drug grand recognition.	No patients failed test ipsilateral to surgery.	No pass/fail criteria presented
Rausch et al. (1984) 6 L-TLE 11 R-TLE 125 mg	Naming, Reading, and Design Matching	Same as Engel et al, above	Longer recovery for Naming and Reading following L injection	Procedure sensitive only to right TLE
DISCRETE ITEM PRESENTATION				
Milner et al. (1962) 50 TLE 200 mg	Two pre- and 2 post-injection line drawings	Recall/recognition of pre-items after speech testing; post-injection items tested after non-specified distractor	With interpretable results, memory failure contralateral to focus in 11 patients	Presence of any error constitute failure for that trial
Milner (1966 & 1972) 123 Ss not all TLE 200 mg	Pre-injection: 2 line drawings, 1 sentence; post-injection: 2 line drawings, 1 rhyme	Same as Milner et al. above	Failure in 25/110 contra injections & 2/116 ipsi injections	Hit rate improved if cases with 1 error not considered; 18 failures, all contralateral to focus

Study/Sample/Dosage	Stimuli	Memory Task	Memory Results	Comment
Kløve et al. (1969) 20 TLE 180 mg	Pre-injection: color & number presentation; post: 3 pictures	Non-specified recall of pre- & post-injection after drug effect	Failure in 4/20 cases with ipsilateral injection	Ipsilateral injections only, No formal pass/fail criteria
Blume et al. (1973) 9 TLE 125+ mg	Pre-injection: color & number presentation; Post: 3 line drawings	Free recall & multiple choice	8/9 patients recalled $\geq$ 2/3rds of line drawings on ipsilateral injection; no post-surgical amnesia in 6 operated cases	Bilateral studies in only 2 patients
Silfvenius & Blom (1984) 3 L-TLE 6 R-TLE 6 Frontal/3 other 150-170 mg	Pre-injection: color, number, sentence, tactual identification Post-injection: 6 words & 6 pictures	Free recall tested during and after drug effect for pre-items; free recall for post-items tested after recovery	Superior performance following non-dominant injection independent of seizure onset	Although no pass/fail criteria, modification of surgery (e.g., stereotactic amygdalotomy, anterior tip resection) based upon memory results
Lesser et al. (1986) 24 L-TLE 12 R-TLE 165 mg	Common objects shown and, if possible, placed in hand following Left injection. Variable number presented immediately after injection	Object recognition following recovery	18/24 L & 4/12 R recognized $\geq$ two-thirds of presented items	Only left hemisphere studies performed
Powell et al. (1987) 14 L-TLE 13 R-TLE Variable	Pre-injection sentence; Post-injection: 3-5 line drawings, 3-5 words, 3-8 objects	Free recall of pre-injection sentence, recognition of post-injection items	10/27 failed and excluded from surgery	Clinical interpretation was subjective without formal criteria

Table 2-3 (continued) Full length reports describing memory performance following amobarbital administration

DISCRETE ITEM STUDIES CONTINUED

Study/Sample/Dosage	Stimuli	Memory Task	Memory Results	Comment
Christianson et al. (1987) 6 L-TLE 7 R-TLE 175 mg	4 concrete words, 4 abstract words, 4 pictures, 4 designs	free recall, cued recall	concrete word memory	Recognition felt more informative
Loring et al. (1989) 40 mixed foci Incremental	2 common objects presented after minimal return of language	aural multiple choice given after return to baseline function	FP errors more frequent following L injection Patients with FP errors displayed poorer baseline neuropsych memory	Relatively few items prevented parametric data analysis
McGlone & McDonald (1989) 13 pts 175 mg	1 object, 2 line drawings, 1 rhyme, and occasionally 1 additional item not specified in report	Recognition memory	7/8 changes in rating from external cause 12/13 repeats yield same results	Combining initial pass/fail ratings inflate accuracy; of initial ipsi memory failures, 1/4 (or 2/5) improved
Rausch et al. (1989) 13 L-TLE 17 R-TLE 125 mg	3 common objects, geometric shape, and unspecified functionally appropriate material	Free recall followed by multiple choice following return to baseline	All patients recognized ≥67% with ipsi injection, 19/30 failed contralateral injection	Temporal lobectomy could be performed on "wrong" side in 11 patients who "passed" contralateral study
Aasly & Silfvenius (1990) 7 L-TLE 8 R-TLE 9 other (6 frontal) 150-175 mg	same pre-inject as Silfvenius & Blom (1984); 3 early & late words, 3 early & late pictures	Post recovery free recall and recognition for pre- and post-injection items	words & pictures poorer after L injection; no difference for early vs. late item performance	Non-parametric data analysis

Study/Sample/Dosage	Stimuli	Memory Task	Memory Results	Comment
Christianson et al. (1990) 12 L-TLE 8 R-TLE 7 L frontal Sz 175 mg	3 concrete & 3 abstract words, 3 pictures, 3 figures, 3 faces	Free recall, cued recall, & recognition	picture recognition best dissociation as fcn of sz focus	Frontal patients included analysis makes comparison of TLE groups difficult
Loring, Lee, Meador et al. (1990) 34 L-TLE 23 R-TLE Incremental	8 objects immediately following hemiplegia; 2 objects, 2 visual spatial designs, and rhyme presented after minimal language return	Recognition following return to baseline	Early & late recognition decreased contralaterally. Lack of agreement between early and late	Hippocampus included in 10 resections despite failing No amnestic syndrome present postoperatively

when there are genuine group differences that have gone undetected. In the present context, a Type I error would indicate that the patient is "at risk" for post-operative memory impairment when no risk genuinely exists, whereas a Type II error would fail to identify appropriately an at risk patient. Type I and Type II errors occur in addition to the 2 conditions in which the correct decision is made (i.e., correctly rejecting or accepting the null hypothesis).

By presenting items soon after administration of amobarbital, the maximal anesthetic effects are present. Thus, to the extent that the hippocampal formation is affected directly or functionally disconnected by the amobarbital injection, hippocampal function soon after injection is most similar to the state that will be created following surgery. However, presenting objects early in the procedure also entails certain risks. For example, medication effects on other cognitive functions are maximal. After left hemisphere injection, global aphasia is usually present for several minutes following injection, and the effect of language failure on memory performance is not fully known. In addition, since patients are frequently akinetic and abulic, one cannot be certain to what degree the items were attended to, although the data of Lesser, Lüders & Morris (1986) suggested that greater attention is present than may be apparent by casual observation. As the medication effects recede, cognitive impairments are less severe, but the presumed effects on the hippocampus are also lessened. Thus, the danger is that performance on items presented later is not sufficiently sensitive in certain patients due to relatively less medication effect on hippocampal function.

In addition to subtotal anesthetization at the time of late item presentation, there is a risk that memory items will be presented at different times following injection of the left and right hemispheres. By waiting to present stimuli until minimal language has returned, items will be presented later following left hemisphere injection than following right hemisphere injection. This problem is not associated with early object presentation.

Retrograde Amnesia

The Wada test has provided valuable information regarding the time course of memory processing. Although in the clinical context the primary concern is whether a subject can learn new information, selective hemispheric anesthesia allows the opportunity to investigate the consolidation of material presented prior to injection. For example, in addition to decreased learning ability, patients who have sustained acute brain injury (e.g., ECT, bilateral temporal lobectomy, head injury, thalamic strokes) frequently demonstrate a retrograde memory loss for material acquired prior to the accident. Despite many epilepsy surgery centers presenting material to be remembered prior to amobarbital injection, there have no reports of either consistent generalized retrograde memory deficits, or material-specific retrograde effects (Rouleau, Labrecque, Saint-Hilaire, Cardu & Giard, 1989; Milner, 1972; Loring, Meador,

Lee & Martin, 1990; Christianson, Silfvenius & Nilsson, 1987; Serafetinides, 1966). Thus, the mechanisms of anterograde and retrograde amnesia likely differ, and the amobarbital effects do not appear to interfere appreciably with the consolidation of newly learned information. Consequently, performance on pre-injection items does not play a prominent part in determining risk for anterograde amnesia.

Methodologic Considerations

Criterion validity refers to the ability of a test to predict a specific event or outcome. However, for criterion validity of this technique to be established in predicting anterograde amnesia, a consecutive series of patients would need to undergo temporal lobectomy without using Wada memory data. If the memory results are themselves used for patient selection, the dependent variable is confounded with the predictor variable requiring validation. When patients who appear to be at risk for post surgical anterograde memory loss are excluded from temporal lobectomy based on Wada results alone, one cannot determine whether the procedure is truly predictive since those very patients believed to be at risk for significant anterograde memory impairment are never operated upon.

Because of the perceived risk of anterograde amnesia, amobarbital memory testing has been used to establish surgery candidacy despite the absence of controlled study. Thus, a consecutive patient series has not been operated upon, and the number of false positive identifications cannot be determined. However, evidence suggests that individuals failing amobarbital memory may not necessarily suffer severe anterograde memory deficits following surgery. Girvin, McGlone, McLachlan & Blume (1987) reported that 3 patients failed the amobarbital memory testing (Montreal approach), but nevertheless underwent standard temporal lobectomy. All 3 patients had bitemporal EEG abnormalities, displayed verbal and non-verbal memory deficits on baseline neuropsychological testing, and had borderline to mildly mentally retarded IQ levels. MR scans revealed left temporal structural lesions in 2 cases. The patients underwent left temporal lobectomy including anterior hippocampus, and none displayed a post-operative amnestic syndrome. Formal followup neuropsychological testing was not presented.

We have also evaluated whether failure on intracarotid amobarbital memory testing would reliably identify individuals at risk for the development of significant post-surgical recent memory deficits (Loring, Lee, Meador, Flanigin, Smith, Figueroa & Martin, 1990). From the series of patients returning for follow-up neuropsychological assessment who had undergone temporal lobectomy including the anterior hippocampus, 10 patients failed either the early or late recognition Wada memory. Failure on early object recognition was operationally defined as corrected recognition scores less than

Table 2-4 Pre- and Post-Temporal Lobectomy Memory Performance for Patients Failing Early or Late Recognition Items

Subj	Dose/Laterality			Buschke	SDL	CF	FSL	Resection
MA	100 mg	L	Pre-	0	0	7	3	6.9 cm
			Post-	40	6	19	6	
JB	100 mg	L	Pre-	40	17	22	19	5.5 cm
			Post-	24	11	21.5	19	
RD	100 mg	L	Pre-	121	20	9.5	1	4.2 cm
			Post-	93	20	14	10	
MJ	175 mg	L	Pre-	17	0	10.5	0	5.2 cm
			Post-	36	--	10.5	1	
AL	250 mg	L	Pre-	30	1	13.5	11	5.8 cm
			Post-	115	16	25	14	
MS	250 mg	L	Pre-	121	22	24	16	5.0 cm
			Post-	52	24	23	16	
JW	150 mg	L	Pre-	38	18	2	8	5.8 cm
			Post-	22	10	9.5	10	
WB	125 mg	R	Pre-	108	21	10.5	8	5.7 cm
			Post-	127	20	14	13.5	
JB	100 mg	R	Pre-	137	22	15	17	5.3 cm
			Post-	130	--	28	--	
CD	100 mg	R	Pre-	103	20	19.5	12	5.0 cm
			Post-	103	24	22.5	18	

Buschke = Selective Remiding Continuous Long-Term Retrieval; SDL = Serial Digit Learning; CF = Complex Figure Memory; FSL = Form Sequence Learning

(Adapted from Loring, Lee, Meador, Flanigin, Smith, Figueroa & Martin, 1990; with permission from *Neurology*)

or equal to 2/8. Due to fewer late items, a corrected recognition score of less than 2/5 was considered failure for the late items. Patients had anterior hippocampus included in the resection because of adequate memory performance during either low-level electrical hippocampal stimulation via intracranial depth electrodes (Loring, Lee, Flanigin, Meador, Smith, Gallagher & King, 1988; Lee, Loring, Smith & Flanigin, 1990), or thermal inactivation during surgery (Flanigin, Schlosberg, Power & Smith, 1985).

Eight patients failed the early object memory testing, 6 patients failed the late item recognition task, with 4 patients failing both early and late memory components. None of these patients developed a dense anterograde amnesia, either subjectively or on formal neuropsychological testing. No patient displayed a decline on all 4 neuropsychological memory tests compared to preoperative levels (see Table 2-4). In fact, most patients attained comparable memory performance levels or improved. It was concluded that failure on intracarotid amobarbital memory testing was by itself insufficient evidence to exclude patients who are otherwise good surgical candidates from temporal lobectomy. Based on failing the additional tests of hippocampal function, 3 additional patients underwent resection of the temporal tip, sparing the hippocampus, and the effect of hippocampal resection in these patients remains unknown. The possibility exists that higher predictive accuracy in our Wada memory study would have been obtained with different memory items, using a different penalty for false positive responses, or presenting items in a forced-choice recognition format. However, because the goal of amobarbital memory testing is to predict a post-surgical amnestic syndrome, and not to predict the extent of material-specific memory deficits, the choice of particular objects should not unduly affect the results. The magnitude of the effect to be predicted far exceeds that which can easily be attributed to methodological variations such as object selection or technique of indicating response. Further, since we employed a technique of dosage determination that differs from the modal approach (Snyder & Novelly, 1991), it could be argued that the technique of serial incremental injections caused a greater number of patients to fail the test than would be observed if a different technique of drug administration had been employed. However, examination of Table 2-1 illustrates that this is not the case. Five of the patients received single bolus injections of 100 mg, which is comparable to the technique and dosage employed in many other centers. Thus, we do feel that a non-representative technique is producing the high misidentification rate.

Novelly & Williamson (1989) reported 25 of 325 patients failed memory testing during the injection ipsilateral to seizure onset. In these 25 patients, a repeat Wada was performed employing a lower medication dosage, and 21 patients demonstrated adequate memory. All 21 patients underwent temporal lobectomy without the development of post-surgical anterograde amnesia. Thus, without repeating the amobarbital memory test, these 21 patients may have been denied the option of surgical treatment.

In their survey of 55 epilepsy surgery centers, Snyder & Novelly (1991) reported considerable variability in amobarbital dosage employed for evaluation, with the majority of programs administering between 100-175 mg. Medication dosage appears to be a critical variable that can affect Wada memory performance. For example, we evaluated amobarbital dosage effects in 29 epilepsy surgery candidates who had adequate bilateral studies and who had undergone subsequent temporal lobectomy (Loring, Meador & Lee, 1991). Since our approach of determining medication dosage was based upon incremental injections necessary to produce a contralateral hemiplegia, medication dosage could be used as a grouping variable in order to investigate possible dosage effects on Wada memory performance. Patients were dichotomized into low (125 mg or lower) and high (150 mg or higher) groups. Following injection ipsilateral to seizure onset, a significant difference between high and low dosages was present (p < .05) for memory performance corrected for false positive errors. In contrast, no differential performance as a function of medication dosage was observed following injection contralateral to the seizure focus. Consequently, medication dosage differences contribute, in part, to the reported differences in memory functioning following amobarbital administration between epilepsy surgery centers.

Psychometric Issues

Rourke & Brown (1986) presented an example of the effects of different base-rates on diagnostic accuracy. The base-rate is the probability that the phenomenon in question exists in the target population and affects the accuracy of clinical decision making. If, for example, the presence of amnesia following temporal lobectomy without selection bias is estimated at 50%, then one-half the patients undergoing resection would be expected to develop a severe anterograde memory deficit following surgery when no predictive procedures are present. Should amobarbital test results correctly classify 90% of at risk patients (correct identification), but also misidentifies 30% of patients not at risk (false positive identification, Type I error), the hit rate will be 80% (correct identification of risk = .9 x .5=45%, correct identification of risk absence = .7 x .5=35%), which is an improvement upon the 50% accuracy expected without testing. However, if the base-rate of memory impairment is 10% with the same diagnostic sensitivity, the diagnostic accuracy would be 72% (correct identification of risk = .9 x .1=9%; correct identification of risk absence = .7 x .9 = 63%). With a base-rate occurrence of only 10%, the diagnostic accuracy in an unselected population with no predictive tests would be 90%. Thus, predictive accuracy actually is reduced when attempts are made at prediction using amobarbital results (72% vs. 90%). In actuality, inflated base-rates of post-resection amnesia have been estimated and the true figure is likely much below 10% (Loring, Meador & Lee, 1990), suggesting even less predictive accuracy.

Reports of Wada memory performance have generally failed to address test reproducibility. Measurement of few samples of transient phenomena is inherently unstable. Dinner, Lüders, Morris, Wyllie & Kramer (1987) repeated the amobarbital memory test in 5 patients who had poor performance following injection ipsilateral to the side of proposed temporal lobectomy. There were significant improvements in performance on the second test. These authors suggested that this degree of performance variability raises concern over the utility of the amobarbital test to evaluate hemispheric memory function.

McGlone & MacDonald (1989) described significant test-retest reliability in patients who received repeat injections due to concerns for either memory or speech function. Of 13 repeat injections, the same pass/fail rating was obtained in 12 cases (p < .02). However, when considering only those patients initially failing the memory test ipsilateral to the seizure focus, a slightly different pattern emerges. Of the 6 patients with deficient memory, 1 patient obtained satisfactory performance on repeat assessment. One additional patient correctly recognized 3/4 target items but also had 2 false positive responses. Based upon "an arbitrary decision . . . concerning . . . passing or failing" (p. 34), this patient was judged to have adequate memory. If a false positive adjustment of total correct minus one-half false positive responses is employed, performance falls below their criterion for passing. Thus, when only clear-cut cases are examined, 1 of 6 patients (17%) initially failing the procedure after ipsilateral injection passed upon reevaluation. If the case with equivocal performance during the initial assessment is included, 2 of 7 patients (29%) obtained satisfactory performance with the second test.

Limitations

Interpretation of memory results may be complicated by inattentiveness, mental confusion, emotional lability, and strong perseverative tendencies (Lee, Loring, Meador, Flanigin & Brooks, 1988; Huh, Meador, Loring, Lee & Brooks, 1989). Memory performance may be also affected by impairment of other "supportive" cognitive functions necessary for the formation of new memories. Testing memory for many types of material is not feasible due to the short duration of drug effect and the medical risk associated with repeated injections. Finally, since the anatomic region of interest, the hippocampal formation, obtains its blood supply from both the internal carotid, and the posterior cerebral arteries, it remains unknown to what degree the hippocampus is affected in each patient tested. Milner (1975) stated that filling of the posterior cerebral artery is not a prerequisite for obtaining memory loss. Thus, the intracarotid amobarbital procedure may test primarily the competency of the neocortical contribution to recent memory acquisition. Others have postulated that perfusion of the septal region of the basal forebrain contributes to the memory impairment following amobarbital perfusion in certain patients (DeToledo, Smith & Kramer, 1989).

Minor cross-flow is commonly present, particularly in younger patients, and the amobarbital procedure can at best only approximate the effects of surgery. Some patients have significant cross-flow of the anterior cerebral arteries, producing a prolonged akinetic/mute state or agitation. Others have normal variations in the cerebral vasculature such as filling of the posterior cerebral artery, making inferences regarding individual patient performances difficult, and different results can be obtained with repeat assessment. Although evidence from SPECT has suggested that angiographic cross-flow does not necessarily indicate functional impairment contralateral to amobarbital injection (Jeffery, Monsein, Szabo, Hart, Fisher, Lesser, Debrun, Gordon, Wagner & Camargo, 1991), this finding is opposite that of Hietala, Silfvenius, Aasly, Olivecrona & Jonsson (1990), who reported frequent contralateral functional impairment.

McGlone & MacDonald (1989) reported that although changes in pass/fail memory ratings were obtained in 8/18 repeat injections, 7/8 changes were associated with identifiable external factors, such as technically unsatisfactory studies, unusual emotional response, or even failing to wear eye glasses. Thus, for many reasons, patients may fail amobarbital memory tests due to factors not directly related to the ability of the contralateral mesial temporal lobe to sustain memory function.

A significant threat to the theoretical model required during Wada memory assessment is the high success rate following injection contralateral to the primary seizure focus. If testing during ipsilateral amobarbital injection is sensitive to temporal lobe dysfunction contralateral to the proposed surgery, as predicted by the model, then memory function also should be sensitive to bilateral lesions (1 physiologic, 1 pharmacologic) when the contralateral side is injected. Impaired memory during injection contralateral to the seizure focus can provide evidence, albeit indirect, supporting the claim of sensitivity to bilateral dysfunction. However, many patients display adequate memory following injection contralateral to a seizure focus. For example, Jones-Gotman (1987) refers to a memory *failure* incidence of only 41% following injection contralateral to a clear, unilateral seizure focus. Although significantly higher than the failure rate of 15% following ipsilateral injection, this figure reveals that nearly 60% of patients performed adequately with known dysfunction contralateral to injection. Rausch, Babb, Engel & Crandall (1989) similarly reported that 11 of 30 patients passed the memory test with adequate performance following injection contralateral to the side of eventual resection. Taken at face value, these results indicate that temporal lobectomy could be performed on the wrong side without concern for an amnestic syndrome in many patients with a unilateral temporal lobe focus. Although a variety of explanations may be proposed to account for these findings, including incomplete anesthesia or atypical vascular distribution, amobarbital memory performance was not affected by injection contralateral to the focus. The high rate of passing following injection contralateral to known dysfunction raises concern regarding this technique's sensitivity to less certain hippocampal dysfunction contralateral to a primary seizure focus.

Posterior Cerebral Artery Injection

Because of the limitations associated with standard intracarotid artery amobarbital injection, a posterior approach through the vertebrobasilar system to anesthetize the medial temporal lobe was developed at the Mayo Clinic by Jack, Nichols, Sharbrough, Marsh & Petersen (1988). In their report, the authors successfully studied 17 patients who were candidates for epilepsy surgery with unilateral injection of the posterior cerebral artery (PCA) ipsilateral to the presumed seizure focus. Following injection, all patients remained oriented and cooperative. Thus, the confounding effects of aphasia on memory function, as well as general attentional deficits, were avoided. No language impairment was present for any of the left seizure patients successfully during the PCA Wada test. All patients had satisfactory memory performance.

Jack, Nichols, Sharbrough, Marsh & Petersen (1988) concluded that given the absence of behavioral confounds, as well as more closely modelling the effects of temporal lobectomy by affecting the hippocampal formation, the PCA Wada technique might develop into a more appropriate method to assess risk for post-surgical amnesia than the conventional carotid technique. However, since the appearance of this report, the PCA Wada test is no longer conducted at the Mayo Clinic (C.R. Jack, personal communication, July 10, 1991). One patient from their series of approximately 50-60 patients undergoing the PCA Wada developed a brainstem stroke as a consequence of this procedure. Even if this risk were only 1%, and their sample size is insufficient to make this determination, the risk-benefit ratio was not felt to be low enough to routinely subject patients to this evaluation in which the risk for post-surgical amnesia is likely less than 1%. The PCA technique offers advantages for cognitive testing, and may have a role in the evaluation of a select subset of patients (e.g., those patients who have a Wada memory test asymmetry in the direction opposite of that predicted by the lesion), although we would recommend a repeat standard carotid Wada prior to a PCA evaluation. The PCA Wada should only be conducted by radiologists who routinely perform interventional neuroradiologic techniques (e.g., AVM embolization).

Conclusions

There is a low base-rate occurrence of significant anterograde memory deficit following temporal lobectomy. Although the issue of post-surgical amnesia is frequently discussed, few reports have appeared in the literature. Further, it appears unlikely that a complete amnesia of the magnitude of H.M.'s amnesia occurs (e.g., the 2 Montreal cases initially reported returned to work). Thus, it may be that performing surgery on a consecutive series of patients would produce amnesia in less than 1% of patients. In addition, advances in evaluation for temporal lobectomy (e.g., intracranial electrodes, more sophisticated neuropsychological assessment, MRI) can better define the areas of seizure onset and cognitive dysfunction, and therefore, help to exclude inappropriate surgical candidates. For example, the autopsy case of Penfield & Mathieson (1974) revealed a pale and shrunken hippocampus contralateral to surgery, and it is possible that this abnormality would have appeared with current MRI scanning protocols (e.g., Press, Amaral & Squire, 1989; Jack, Bentley, Twomey & Zinsmeister, 1990; Jack, Sharbrough, Twomey, Cascino, Hirschorn, Marsh, Zinsmeister & Scheithauer, 1990).

The effects of post-surgical memory impairment may be considered so devastating to a patient that an epilepsy surgery center may wish to eliminate all possibility of post-resection memory deficit. The simplest way would be to stop performing temporal lobectomy. A center also could choose to operate only on right handed/left cerebral dominant patients with seizures clearly originating from the right mesial temporal lobe. However, many people with left seizure onset would be denied treatment. Most centers include left seizure patients if there is a strictly unilateral seizure onset, although the 3 cases of global amnesia in the Montreal series occurred after temporal lobe resection of the language dominant hemisphere (Milner, 1972). At present, each center, in consultation with the patients undergoing evaluation, decides what are acceptable risks prior to undergoing surgery. Some patients are willing to accept the risks because they feel that life with frequent debilitating seizures is without significant quality. In this context, 1 patient in our series who underwent temporal lobectomy at the Medical College of Georgia developed a post-operative worsening of her pre-operative recent memory deficit. It precluded her from returning to work, and she was unable to remember how to get around in her hometown. However, she became seizure free following her surgery. Although well aware of the limitations imposed by her memory deficit, she and her family repeatedly stated that they would again choose surgery knowing its effects upon the patient's memory. Thus, an individual's weighing of the cost/benefit ratio needs to be considered.

Practical Considerations. In the clinical evaluation for epilepsy surgery, we interpret poor amobarbital results as suggesting possible risk to recent memory, and conclude that additional studies of hippocampal function will be needed. When a strong amobarbital recognition asymmetry, as predicted based on known seizure onset, is observed (e.g., 5/8 or more early objects with injection ipsilateral to seizure onset and 2/8 or fewer contralaterally), we infer that surgery including the hippocampus will not produce a serious anterograde amnesia. Thus, the test identifies low-risk patients rather than accurately predicting those with potential post-resection memory difficulty, although such an asymmetry adds supportive evidence for lateralization of the seizure focus.

In one patient who was evaluated for epilepsy surgery at the Medical College of Georgia, we observed exclusive right hemisphere language representation, and baseline neuropsychological evaluation indicated relative verbal memory impairment. An asymmetry in early object memory was present, with 8/8 objects recalled following left hemisphere injection, but 0/8 objects recalled following right hemisphere injection. Recall was normal for all items presented later in the procedure (3/5 items with no false positive responses), and no left/right difference was observed. Despite depth electrode implantation, seizures could not be localized with confidence, and a decision on his surgery was deferred. However, the family pursued surgery at a different center. The patient was re-evaluated with invasive EEG procedures and his typical spontaneous seizures were observed to originate from the right temporal lobe. A right temporal lobectomy was performed. Following surgery, a significant language deficit was noted. In addition, a profound recent memory deficit was observed, and this has persisted for approximately 1 year (decrease in WMS-Revised, General Memory Index from 84 to 58). Although hindsight is 20/20, given the large Wada memory asymmetry on early objects, we would have repeated the Wada memory test with a smaller medication dosage, and if the same memory asymmetry was present, chosen to perform hippocampal cooling studies prior to right hippocampal resection. Only the temporal tip would have been resected unless there was strong evidence at the time of surgery to suggest that the hippocampus could be resected safely. However, this pattern of strong asymmetry in the opposite direction of that predicted is a different kind of evidence from that in which bilaterally poor memory is obtained. This case also illustrates the differential sensitivity of early and late item presentation.

When poor memory is obtained following amobarbital injection ipsilateral to proposed surgery, several options are available depending on the biases of the surgical institution in which the patient is being evaluated. The patient may be excluded from surgery. However, this may needlessly exclude candidates who can benefit from the surgery. For example, Powell, Polkey & Canavan (1987) excluded 37% of their sample because they felt that poor amobarbital memory results contraindicated surgical intervention. Patients at the Montreal Neurological Institute who fail this test, yet are otherwise good surgical candidates, undergo a limited resection in which the hippocampus and hippocampal gyrus are spared (Jones-Gotman, 1987), although there is a decreased likelihood of good seizure control.

Some institutions, including our own, repeat the amobarbital memory test at the same or lower dose when patients fail the initial test (e.g., Novelly & Williamson, 1989). If adequate performance is obtained upon repeated assessment, then temporal lobectomy can be performed. Alternatively, special memory testing may be conducted such as with a specialized posterior amobarbital technique (e.g., Jack, Nichols, Sharbrough, Marsh & Petersen, 1988). Similar criticisms regarding the validity of the posterior Wada studies can be advanced. However, if each technique falsely identifies 10% of patients as at risk, and passing a single test is sufficient to proceed to surgery involving the hippocampus, then repeating the standard Wada test when necessary will result in false identification of risk in only 1/100 cases (i.e., 0.1 x 0.1). If the PCA Wada test is then performed following 2 standard carotid tests, then only

1/1000 patients will be misidentified using the same figures (i.e., 0.1 x 0.1 x 0.1).

If there is sufficient evidence suggesting a unilateral seizure onset despite poor amobarbital memory, such as with repeated seizure recordings, clear asymmetry in metabolic activity (PET or SPECT), or the presence of a structural lesion, a temporal lobectomy including hippocampus may be considered in certain cases and the amobarbital memory results discounted (e.g., Girvin, McGlone, McLachlan & Blume, 1987). We suggest that the Wada memory results not be considered as absolute, and should be interpreted within the entire clinical context of pre-operative epilepsy surgery evaluation.

3
EEG and Neurologic Functions

The use of EEG during intracarotid amobarbital testing varies from center to center (Rausch, 1987). When used, its primary purpose is to provide a continuous physiological measure of the effect of amobarbital on cortical function. In order to understand the use of EEG during this procedure, it is helpful to review some of the major studies of the neurological and EEG effects of intracarotid amobarbital. This chapter will provide a review of these studies and a brief summary of the present use of EEG during the Wada test.

Review of Major Studies

Werman, Christoff & Anderson were among the first investigators to systematically study the effects of intracarotid amobarbital sodium on behavior and EEG (Werman, Anderson & Christoff, 1959; Werman, Christoff & Anderson, 1959). They studied 16 patients undergoing carotid arteriogram for various neurological causes. All had normal or mildly abnormal baseline EEGs, and all had carotid arteriography which showed normal intracarotid vasculature. From 25-75 mg of amobarbital sodium were injected into the internal carotid artery. All but 1 patient showed neurological changes within 5 seconds of injection, and the effects lasted up to 7 minutes. Neurological changes included contralateral hemiparesis, contralateral hemisensory deficit, and contralateral hemianopsia. Aphasia was not observed regardless of side injected. When larger doses were used, "diffuse" effects, such as nystagmus, dysarthria, and sleep were noted. Fourteen patients also underwent the intravenous injection of amobarbital. All developed nystagmus and dysarthria, but only 3 showed hemisphere signs.

In order to assure that the EEG effects were caused by the initial pass of the injected amobarbital, these investigators studied the EEG immediately following the injection. They reported that EEG changes occurred within 4 seconds of injection. EEG changes included bursts of diffuse high voltage slow activity usually lasting less than 30 seconds despite the fact that behavioral changes lasted up to 7 minutes. Although slowing was present bilaterally, it was often more prominent over the ipsilateral side. Data on individual subjects were not provided. These authors concluded that the intracarotid injection of amobarbital resulted in unilateral behavioral effects, but bilateral EEG effects during the first 30 seconds following injection. The results were not related to pathology (Werman, Anderson & Christoff, 1959; Werman, Christoff & Anderson, 1959).

The classical study of intracarotid amobarbital for lateralization of speech dominance was published by Wada & Rasmussen (1960). It included experimental studies on monkeys and clinical studies on patients being evaluated for epilepsy surgery. The investigators carried out 15 injections of 7-350 mg of amobarbital into the right common carotid artery of 6 monkeys. All injections resulted in a contralateral hemiplegia, and doses greater than 100 mg resulted in anesthesia. Small doses resulted in unilateral, short duration EEG effects ipsilateral to the side of injection. Larger doses resulted in more generalized and prolonged EEG effects. Amobarbital was also injected into the right vertebral artery of 5 monkeys. Within 1-2 seconds of injection, all animals became flaccid and were anesthetized. With smaller doses, anesthesia occurred within 1-2 minutes. EEG changes were generalized.

As a part of this same study, Wada & Rasmussen (1960) reported 20 patients who underwent Wada testing as a part of their evaluation for epilepsy surgery. The left and right common carotid arteries were studied in each patient on separate days with 150-200 mg of amobarbital, which was injected over 1-2 seconds. There was an immediate flaccid hemiplegia contralateral to the side of injection. Regardless of side of injection, there was an initial period of "confusion," hesitation, and cessation of counting lasting for a few seconds. If the injection occurred on the dominant side, the cessation of counting persisted as a part of a global aphasia. Despite a severe contralateral hemiplegia, if the injection were on the nondominant side, the patients were able to resume counting and language function within 5-20 seconds. The duration of the motor and language impairment varied with the dose. In general, severe hemiplegia and language disturbance persisted for 1 1/2 to 5 minutes followed by a gradual return over an additional 1-5 minutes. These results, quite different from those of Werman and associates, conclusively demonstrated that aphasia resulted from injection of amobarbital into the carotid artery supplying the dominant hemisphere. EEG results on these patients were not included in this report (Wada & Rasmussen, 1960).

Perria, Rosadini & Rossi (1961) reported behavioral and EEG effects of amobarbital in 30 patients with cerebral neoplasms, seizures, or cerebral vascular disease undergoing carotid angiography. Nineteen patients were injected unilaterally, and 11 were injected bilaterally. Injections were made into the common carotid artery or the proximal portion of the internal carotid artery over 4-6 seconds. From 2.5 to 100 mg of amobarbital were injected. They found 2 types of effects: those which occurred independent of side of injection and those which were related to hemisphere specialization. Those independent of side of injection included EEG changes, effects on motor and reflex function, and an initial cessation of verbal communication. Effects which were associated with hemisphere specialization included effects on language function and emotional reactions.

EEG changes occurred within 4-6 seconds after the beginning of the injection. Injections of 5-10 mg resulted in the development of low amplitude beta activity, and injections up to 100 mg resulted in high amplitude delta activity. Intermediate doses resulted in a mixture of beta and delta activity. The duration of EEG changes varied with the dose, i.e. the effects of smaller doses disappeared more rapidly, and the effects of larger doses required at least 5 minutes for disappearance. Following the larger doses, the sequence of return was as follows: high amplitude slow waves, followed by a mixture of slow waves and beta activity, followed by low amplitude beta activity, and finally a return to baseline activity. Thirty-seven of 41 injections showed unilaterally predominant EEG effects ipsilateral to the side of injection. Four injections resulted in definite bilateral EEG changes. The bilateral changes did not appear to be related to the dose or to obvious arterial abnormalities.

Effects on motor and reflex function included contralateral hemiparesis, contralateral hyperreflexia, and a Babinski sign in most patients. These occurred if the dose of amobarbital was at least 50 mg. Three patients demonstrated prominent bilateral motor effects, and these were 3 of the 4 patients who showed bilateral EEG changes. Regardless of side of injection, most patients stopped counting and were unable to execute simple movements for up to 30 seconds. The authors considered this brief period of speech cessation to represent a transient impairment of consciousness.

Effects which were dependent on hemisphere specialization included effects on language function and emotional reactions. Complete aphasia occurred only after doses of at least 100 mg. It was simultaneous with maximal EEG, motor, and reflex changes. In 2 cases, a language disturbance occurred following injection of either side, and these were 2 of the patients who demonstrated bilateral EEG changes. Some patients demonstrated "depression" when the language dominant side was injected, and euphoria often occurred when the nondominant side was injected. These reactions tended to occur 4-6 minutes following the injection and lasted for 1-10 minutes. The 4 patients who showed bilateral EEG changes did not demonstrate this type of reaction, and a number of patients had no reaction even with unilateral EEG changes.

Four patients also received an intravenous injection of 100 mg of amobarbital. The IV injection of 100 mg resulted in no EEG change and no clinical disturbance. Four hundred mg of amobarbital was required intravenously before patients developed generalized beta activity and drowsiness. As a result, these authors concluded that recirculation of amobarbital was unlikely to be of importance in standard Wada testing. They further concluded that EEG, motor, and reflex changes could be used to document that the amount of barbiturate injected was adequate to determine if language representation was present on the injected side (Perria, Rosadini & Rossi, 1961).

In 1965, Serafetinides and colleagues reported 21 patients who underwent Wada testing (Serafetinides, Driver & Hoare, 1965; Serafetinides, Hoare & Driver, 1965). Three patients underwent unilateral injections, and 18 had bilateral injections. In the first 5 patients, 125 mg of amobarbital was rapidly injected into the internal carotid artery. Because consciousness became impaired following 6 of the 10 injections, the later patients underwent a slower injection. These investigators found that of 8 right handed patients with left hemisphere dominance for language, 7 lost consciousness when the dominant hemisphere was injected, and 1 showed minimally impaired consciousness when the nondominant hemisphere was injected. Of 7 patients who were left handed or ambidextrous but who had definite left hemisphere dominance for language, 6 lost consciousness when the dominant hemisphere was injected. Two patients developed slightly impaired consciousness when the nondominant hemisphere was injected. Three patients demonstrated bilateral speech representation, all of whom lost consciousness when either side was injected. EEG slowing in all of these patients was confined to the side injected. This was despite the fact that consciousness became impaired only when the dominant hemisphere was injected. These authors concluded that the dominant hemisphere was predominant over the nondominant hemisphere with respect to consciousness (Serafetinides, Driver & Hoare, 1965; Serafetinides, Hoare & Driver, 1965).

In 1967, Rosadini & Rossi reported their results in 69 patients. They attempted to confirm the studies of Serafetinides and colleagues which had suggested that consciousness is a function of hemisphere specialization. Patients included were those with epilepsy, vascular disease, neoplasms, and psychiatric disorders. In 48 studies in 38 patients, awareness was tested by examining the capacity of the patient to keep in contact with the examiner immediately following the injection. In 21 studies in 14 patients, the patient was instructed to manipulate a switch with the nonparetic hand following a visual or auditory stimulus. Amobarbital (100-200 mg) was injected over 4-5 seconds into either the common carotid artery or the proximal internal carotid artery. Patients were tested for motor function, speech function, and EEG. Of the 48 injections during which consciousness was studied by standard clinical examination, there was no sign of impairment of consciousness following 31 injections. All of these patients showed ipsilateral EEG changes and contralateral hemiparesis. Sixteen of these 31 patients developed an aphasia, and 15 developed no aphasia. Nine of the 48 injections resulted in a transitory depression of communication lasting less than 1 minute. Five were from the left side, and 4 from the right side. In 5 of these patients, some degree of bilateral slowing was present on EEG, although the slowing was maximal ipsilaterally. Following 8 of the 48 injections, impaired consciousness lasted more than 1 minute. Five injections were on the dominant side, and 3 were on the nondominant side. In 7 of these cases, the injection was contralateral to an occluded carotid artery and a damaged hemisphere. In these 7 patients, the injections produced bilateral EEG slowing.

Following 21 injections, patients were instructed to manipulate a switch following a visual or auditory stimulus. There was no impairment of performance following 16 of the injections, 8 on the right side and 8 on the left side. All 16 patients developed ipsilateral EEG slowing and contralateral hemiparesis, and the 8 left-sided injections resulted in aphasia. These patients did not fail to operate the switch during the entire test. One of the 21 patients demonstrated a transitory arrest of function for approximately 30 seconds after a right-sided injection. The EEG showed bilateral slowing more marked ipsilaterally even though the hemiparesis was unilateral. Four patients demonstrated a prolonged impairment of switch operation. Two had contralateral carotid occlusions, and 1 had manual compression of the contralateral carotid artery. All 3 developed bilateral EEG slowing and bilateral motor impairment suggesting a bilateral effect of the amobarbital. Angiography showed bilateral distribution of the contrast material in these 3 patients. In the final patient, a left carotid injection resulted in prolonged impairment of consciousness. The patient had large left posterior communicating and left posterior cerebral arteries, and the impaired consciousness may have been secondary to involvement of posterior fossa structures. The EEG showed only mild slowing on the ipsilateral side.

The authors concluded that prolonged impairment of consciousness following amobarbital injection was secondary to functional impairment of a large part of both hemispheres or of one entire hemisphere. They also concluded that transient impairment of function is uncommon but does occur, and this likely represents a brief shift of attention following the injection. There was no evidence of a hemisphere dominance for consciousness (Rosadini & Rossi, 1967).

Kløve and colleagues in 1969 reported their results in 20 patients who underwent Wada testing. They injected 180 mg of amobarbital over a 4 second period. All patients demonstrated unilateral EEG slowing affecting primarily the anterior 2/3 of the hemisphere for up to 7 minutes following the injection. On occasion, there was slight contralateral frontal slowing. They found that beta activity was present ipsilaterally initially and that the beta activity later spread to become bilateral (Kløve, Grabow & Trites, 1969).

Rausch, Fedio, Ary, Engel & Crandall (1984) studied the recovery rates of various neurological functions and the EEG following intracarotid amobarbital injections. They studied 17 patients with unilateral temporal epileptiform foci, 6 on the left side and 11 on the right side. All patients were left hemisphere dominant for language. An injection of 125 mg of sodium amobarbital was administered over 4-6 seconds, and testing began approximately 20 seconds post injection. All patients demonstrated a transient aphasia following left hemisphere injection. The investigators found that the duration of ipsilateral EEG slowing was not affected by the patient group or the hemisphere injected independently. On the other hand, taking both patient group and side of injection together, significant differences were present. Patients with left temporal lobe seizures showed more prolonged ipsilateral slowing following injection of the right than left hemisphere, and patients with right temporal lobe seizures showed more prolonged ipsilateral slowing following left than right injection. They hypothesized that the epileptic hemisphere may interfere with recovery of the contralateral "normal" hemisphere, i.e. there may be a negative functional effect of the epileptic hemisphere on the contralateral normal hemisphere (Rausch, Fedio, Ary, Engel & Crandall, 1984).

Lesser, Dinner, Lüders & Morris (1986) studied 36 patients with complex partial seizures to determine if memories could be formed during the initial stage of mutism and apparent confusion following amobarbital injection. They found that new memories could be formed during this initial period. Their data suggested that the initial period of mutism was not related to cross flow or bilateral EEG changes. Unlike Rausch, Fedio, Ary, Engel & Crandall (1986), they found that the duration of EEG slowing following intracarotid amytal injection was not affected by either the side of injection or the side of epilepsy (Lesser, Dinner, Lüders & Morris, 1986).

Neurological and EEG Effects

As usually performed, both neurological and EEG effects begin immediately following amobarbital injection. Within seconds, the patient develops a contralateral hemiparesis, contralateral hemisensory deficit, and frequently a contralateral hemianopsia. In some patients, there appears to be a brief period of mutism, confusion, or inattention lasting less than 30 seconds regardless of the side of injection. If the injection is on the dominant side, the patient becomes aphasic and is thus unable to resume counting. In the usual case, the hemiparesis and other neurological effects last for 5-10 minutes, and there is a gradual return to baseline function. The mechanism of the initial period of mutism and/or inattention is poorly understood. As pointed out by Lesser, Dinner, Lüders & Morris (1986), patients are able to form new memories during this initial period of time.

Within seconds of the amobarbital injection, high amplitude delta activity develops over the ipsilateral hemisphere, primarily over the frontal and temporal regions. There is often less prominent contralateral slowing, and when present, it is usually most prominent over the contralateral frontal region. These EEG changes persist for 5-10 minutes following the injection and correspond fairly well to the period of neurological deficit. The return of EEG activity from delta activity to baseline may go through a stage of prominent beta activity similar to that seen during the injection of a low dose of amobarbital. There are data suggesting that the brief period of confusion/mutism/inattention immediately following injection may be associated with bilateral EEG slowing. There are also data that some patients with this abnormality have abnormal circulation resulting in bilateral effects of the intracarotid amobarbital.

A survey of 15 centers participating in the International Conference on the Surgical Treatment of the Epilepsies revealed that 40% of the centers routinely performed EEG during the Wada examination, 27% sometimes performed EEG, and 33% never used EEG as a part of their testing (Rausch, 1989). In a larger survey of 55 centers, Snyder & Novelly (1990) found comparable results, with 51% of the centers always obtaining concurrent scalp EEG, 29% recording EEG at least some of the time, and 18% of the centers never recording EEG. Thirty-four percent of the centers stated that EEG results "always" affected interpretation of the Wada data, whereas 40% of the respondents reported that EEG results affected the data interpretation at least some of the time.

In general, the primary purpose of performing an EEG during the Wada examination is to document the effect of amobarbital on cortical function. Ipsilateral or predominantly ipsilateral delta activity suggests that the amobarbital has affected only the ipsilateral cerebral hemisphere. When unilateral slowing can be demonstrated soon after the injection, testing for language function and the presentation of items for memory testing can begin. If bilateral slowing is present initially, some centers wait until the bilateral effects subside before beginning the testing. As noted above, unilateral slowing tends to persist for 5-10 minutes following injection and correlates fairly well with the period of neurological deficit. The disappearance of unilateral slowing along with the return of neurological functions suggests that the hemisphere has returned to normal. Thus, when unilateral slowing has disappeared and baseline activity has returned, testing may begin for recall and/or recognition of the items presented during the period of unilateral hemisphere dysfunction. Persistent bilateral slowing may suggest intracranial pathology or abnormal circulation. When bilateral slowing persists, the results of the language and memory testing must be interpreted with caution.

Summary and Conclusions

The major effects of intracarotid amobarbital on the EEG have been well documented and confirmed by numerous investigators. The primary effect from standard doses of intracarotid amobarbital is ipsilateral slowing over the frontal and temporal regions with occasional milder slowing over the contralateral frontal region. This effect correlates well with the unilateral behavioral effects associated with intracarotid amobarbital. The EEG is used by a number of centers during the Wada evaluation as a physiological measure of cortical function. It may assist in determining when it is appropriate to begin language testing and to present items for memory testing and when to begin the recall/recognition portion of the testing. Unusual EEG changes may lead one to question whether abnormal circulation is present and whether the test can be used as a valid measure of unilateral hemisphere language or memory function.

4
Attention

Wada testing was developed to assess cerebral lateralization of language and has been employed to evaluate memory during the presurgical evaluation of temporal lobe epilepsy. However, unilateral intracarotid amobarbital injection produces a wide range of extralinguistic and extramnemonic neurobehavioral disturbances including contralateral hemiplegia, hemisensory loss, extinction, homonymous hemianopsia, changes in color vision, anosognosia, anosodiaphoria, asomatognosia, euphoria, depressive-catastrophic reactions, confusion, and loss of consciousness (Terzian, 1964). The integrity of attentional mechanisms during the intracarotid amobarbital test has been generally ignored, even though possible alterations in these mechanisms may confound the interpretation of test results.

Although intracarotid amobarbital injection can potentially confound the interpretation of clinical results, the Wada test offers a unique opportunity to investigate differential effects for symmetric acute dysfunction of each cerebral hemisphere in each subject. For example, the Wada test has been employed to demonstrate right cerebral dominance for a specialized role in directed attention toward the external milieu. The effect of intracarotid amobarbital on the content of consciousness is less clear, especially following left cerebral injections. In order to further delineate differential hemispheric contributions to the multiple facets of attentional behavior, novel experimental paradigms and an increase in knowledge of the anatomic and physiologic substrate will be required. In this chapter, the effects of intracarotid amobarbital on attentional mechanisms will be reviewed, including issues related to consciousness, vigilance, and specialized roles for the left/right cerebral hemispheres.

Arousal and Consciousness

Consciousness is the state of awareness of self and the environment (Plum & Posner, 1983). The precise limits of consciousness are difficult to define because we can only infer self-awareness in others by their appearances or actions. Consciousness involves both the content of mental functions and arousal, which is closely linked to wakefulness. A related concept is attention, which has a variety of definitions. We shall define attention as a state of readiness to perceive or act on a stimulus or a range of stimuli (Wolman, 1973).

Behavioral changes suggesting an impairment in arousal or consciousness following intracarotid amobarbital injection have been mentioned by many authors (Wada & Rasmussen, 1960; Gilman, MacFayden & Denny-Brown, 1963; Terzian, 1964). Further, Serafetinides, Hoare & Driver (1965) reported a relationship between consciousness and hemispheric dominance for language. They examined the effects of intracarotid amobarbital (125 mg) on consciousness as assessed clinically in 21 patients, 18 of whom received bilateral injections. Of the 8 dextral patients who received bilateral studies, 7 lost consciousness only after injection of the language dominant side. In the 1 other dextral patient, consciousness was lost after each injection, but the duration of unconsciousness was shorter on the nondominant side. In 7 sinistral or ambidextrous patients with unilateral language dominance, unconsciousness followed injection of the dominant side on 6 occasions, and of the nondominant side on 2 occasions, but again unconsciousness was of shorter duration on the non-dominant side. In the 3 patients with mixed language dominance, consciousness was impaired after injection on each side. Serafetinides, Hoare & Driver (1965) concluded that consciousness is linked to a greater degree with function of the cerebral hemisphere dominant for language.

However, other authors have reported that consciousness is not suppressed following unilateral injection into either hemisphere. Rosadini & Rossi (1967) examined a total of 69 amobarbital injections (usually 100-200 mg) in 52 patients. In their first study, they assessed consciousness based on "clinical semeiology" following a total of 48 injections in 38 subjects (i.e., only 10 received bilateral injections). In their second study, they quantified the subjects' consciousness and capacity to respond by having them press a hand switch held ipsilateral to the intracarotid injection whenever they heard "a given sound or saw a flash of light." Fourteen patients were examined in this manner; 7 received bilateral examinations and 7 were tested only on 1 side. Out of the total 69 injections, "no sign indicating the occurrence of loss of consciousness" was found after 47 injections. About half of these involved the language dominant hemisphere. A brief impairment of consciousness lasting less than 1 minute was noted after 10 injections (5 on the dominant and 5 on the non-dominant side). Twelve injections produced clear loss of consciousness for greater than 1 minute (8 on the dominant side and 4 on the non-dominant side). In the subset of patients receiving the stimulus-response test (21 injections in 14 patients), no impairment was noted following 16 intracarotid injections, with the patients never failing to operate the switch in response to the signal during the whole duration of the test. A transient arrest was noted in 1 case and long-lasting suppression of switch operation was recorded after 4 injections (3 left and 1 right). In all cases with loss of consciousness, Rosadini & Rossi (1964) felt that the occurrence could be explained by contralateral severe brain damage or by crossflow to the contralateral hemisphere or posterior circulation.

Fedio & Weinberg (1971) also reported a failure to confirm a lateralized hemispheric relationship between consciousness and language mechanisms. In 12 epileptic patients (8 dextral, 4 sinistral), they compared the effects of left and right amobarbital injections on a continuous sequential naming and recall task. Serial projections of pictures were alternated with the word "AND." Subjects were instructed to identify the word "AND" whenever presented, and to name each picture as presented and recall the prior picture. Following each intracarotid injection, recovery as indicated by the first correct response usually was in the order of "AND," naming, and finally recall. Overall, recovery was selectively slowed for the left hemisphere. Average recovery time and range (in minutes) for left/right injections were as follows: AND = 0.5(0.1-1.6) left, 0.2 (0.1-.9) right; naming = 1.1 (0.1-2.4) left, 0.3 (0.1-1.2) right; recall = 2.9 (0.5-4.7) left, 0.7 (0.2-2.0) right. Significant left-right differences were noted for recovery of naming and recall. The left-right difference to initiation of speech (i.e., saying "AND" in response to the slide) was not statistically significant.

Medical College of Georgia Study. In order to re-assess the effects of unilateral intracarotid amobarbital injection on attentional mechanisms, we undertook a prospective study (Huh, Meador, Loring, Lee & Brooks, 1989), examining the hemispheric effects of intracarotid amobarbital on a simple continuous performance task similar to the task employed by Rosadini & Rossi (1967). Subjects were instructed to respond to a randomly occurring strobe light flash by depressing a hand-held button once for each flash with the hand ipsilateral to side of injection. The strobe light was positioned 12 inches from the head ipsilateral to side of injection. The flash stimuli and responses were recorded with an EEG machine. Twenty-three consecutive patients undergoing bilateral intracarotid amobarbital tests as a part of their preoperative evaluation for epilepsy surgery were included in the study. Demographics were as follows: sex = 13 men, 10 women; mean age = 30 years (range 18-55); handedness = 20R, 3L; language dominance = 17 L, 1 R, 5 mixed; and seizure focus = 11 left mesial temporal lobe (MTL), 7 right MTL, 2 right frontal lobe, 3 unlocalized. All patients received incremental injections of amobarbital (initial dosage 100 mg, 50 mg increments) necessary to produce contralateral flaccid hemiplegia. Mean left hemisphere dosage was 122 mg, mean right side dosage was 128 mg. Injections were administered by hand, and the order of injection was sequentially alternated. Strobe light flashes were delivered at random intervals but were controlled to avoid simultaneous exposure with other test items. Care was also taken to deliver flash stimuli when the subject's eyes were open; if severe ptosis was present, the lids were manually elevated during periods of stimulus presentation. Responses occurring within 1.2 seconds following flash stimuli were considered correct. This time period was chosen to avoid excluding slowed correct responses. We regarded more than 1 response for each light flash and all other inappropriate hand-button responses as perseverative or false positive errors. Number of stimuli, correct responses and perseverations were counted during the initial 180 seconds following injection. The mean number of flash stimuli was 17 for preinjection trials and 36 following unilateral amobarbital injections. For comparison of individual performance following each injection, 3 indices were calculated for each of the 46 injections: response index = (no. responses/no. stimuli) X 100; percent correct index = (no. correct responses/no. stimuli) X 100; percent perseveration index = (response index - percent correct index).

Prior to injection, subjects performed the task well bilaterally. In contrast, task performance was markedly altered following each intracarotid amobarbital injection in all subjects (see Table 4-1). Attention was severely impaired following all amobarbital injections as noted by the decrease in correct responses for both the right and the left amobarbital tests ($p < .0001$ each). All subjects failed to respond properly to a large number of flash stimuli. In addition, unilateral intracarotid amobarbital injection produced a significant increase in perseverative responses ($p < .0002$ right injection; $p < .004$ left injection).

We compared task performances following right and left hemispheric injection to examine possible differential hemispheric effects on attentional mechanisms. Higher responsiveness ($p < .026$) and more correct responses ($p = < .017$) occurred with right as compared to left intracarotid amobarbital injections. Perseverative responses were not significantly different. The left/right asymmetry in response was not present in all subjects. A left/right response index asymmetry $\geq$ 20% was present in 17 of the 23 subjects. In 11 subjects, responsiveness was less for left than right intracarotid amobarbital injection; 6 subjects had less responsiveness following right injection; the remaining 6 subjects demonstrated less than 20% asymmetry. None of these asymmetries were explained by seizure focus, handedness, language dominance, or order of injection. However, contralateral whole hemisphere angiographic crossflow was greater following left than right injection ($p < .02$).

Independent of seizure focus, perseverations reduced the subsequent memory recognition score ($p < .05$). Further, in patients with left/right asymmetries in both memory score and perseverations, the memory score was less on the side with greater perseverations in 14 of 15 patients ($p < .0005$). Similar analyses comparing response index and percent correct index to memory score were not significant.

In the subjects with a unilateral mesial temporal lobe seizure focus, the

Table 4-1 Comparison of task performances: Preinjection, post-left injection, post-right injection, and combined post-left and -right intracarotid amobarbital median index scores

	Response Index	Percent Correct	Percent Perseveration
Preinjection	100	100	0
Left	35	7	12
Right	85	24	30
Left and right	41	19	16

(From Huh, Meador, Loring, Lee & Brooks, 1989; reproduced with permission of *Neurology*)

focus vs. nonfocus sides were compared for memory score, response index, perseveration index, and percent correct index. Only memory score was significantly different between the focus and non-focus sides ($p < .03$). However, the correct asymmetry was present in only 10 of the 18 patients with unilateral mesial temporal lobe focus. Response index, perseveration index and percent correct index were examined as a function of memory score for the focus and nonfocus sides. Memory score on the nonfocus side was significantly correlated with the response index ($p < .007$) and with the number of perseverations ($p < .0002$); no other correlations were significant. Thus, an increase in perseverations on the nonfocus side was associated with poorer memory performance.

Conclusions. Unilateral intracarotid injection of amobarbital produces considerable negligence of response to stimuli. Consciousness varies along a continuum, and there is no sharp distinction between unconsciousness and fully intact consciousness. From our observations, we do not think that consciousness is fully intact, although some aspects of consciousness remain operative following unilateral intracarotid amobarbital injections. For example, language function may remain intact following nondominant hemispheric injections. Further, objects presented after amobarbital injection may be recognized after resolution of amobarbital effects (Lesser, Dinner, Lüders & Morris, 1986). However, subjects frequently exhibit impaired free recall of events during the first few minutes following intracarotid amobarbital injection. In addition, they frequently have ptosis and may exhibit decreased responsiveness to pain, decreased spontaneous movements, and at times yawning. Even late in the intracarotid amobarbital procedure (i.e., >6 minutes post injection), attention for tactile stimuli is impaired (Meador, Loring, Lee, Brooks, Thompson, Thompson & Heilman, 1988). Thus, arousal is at least partially impaired by unilateral intracarotid amobarbital. The effect of intracarotid amobarbital on the content of consciousness is less clear, especially following left intracarotid injections.

The deficits observed in our study are probably multifactorial. Adequate performance of the continuous performance task requires intact sensory mechanisms for recognition of relevant stimuli, temporal processing, and prompt connection to motor programs for execution of appropriate behavioral responses. Therefore, interruption of 1 or more of these components would result in decreased performance. Since the visual stimuli were delivered to the hemifield ipsilateral to the amobarbital injection and the required response was by the hand ipsilateral to the amobarbital injection, the reduced responsiveness cannot be attributed to an impairment of primary motor or sensory systems. However, dysfunction of executive mechanisms could produce a failure to initiate responses (i.e., akinesia) or produce a dissociation of the sensory and motor aspects of attentional behavior, resulting in an uninhibited release of pre-existing motor programs, i.e., perseverations (Sandson & Albert, 1987). A confusional state cannot be ruled out in the setting of aphasia but does not explain performance deficits for the majority of the nondominant hemispheric injections. Similarly, a mnestic disorder cannot be ruled out in every case, but seems an unlikely explanation for the perseverative response errors. Further, during the right side amobarbital test, many patients can remember items presented prior to injection. Finally, the obvious reduction in arousal suggests that at least a portion of the performance deficits must be related to impaired attentional mechanisms.

We observed a greater impairment of the attentional task following left intracarotid amobarbital. However, this asymmetry may be due to left/right differences in crossflow. Impairment of attentional behavior following unilateral intracarotid amobarbital may result from alterations of lateralized function unique to each hemisphere. Previous studies have shown a right cerebral dominance for spatial and tactile attention (Meador, Loring, Lee, Brooks, Thompson, Thompson & Heilman, 1988; Weintraub & Mesulam, 1987). Selection and execution of an appropriate motor program in the presence of conflicting motor signals is analogous to selection of sensory stimuli in the spatial domain. The left hemisphere is believed to contain motor engrams that are necessary for elaborating skilled purposeful movement (Heilman, Gonzales & Rothi, 1985). Therefore, one can speculate that right intracarotid amobarbital would produce failure in sensory scanning and subsequent release of motor inhibition, whereas left intracarotid amobarbital would interfere primarily with the executive mechanisms of motor attention required to perform the task (Huh, Meador, Loring, Lee & Brooks, 1989).

Some of the observed alterations in attention may be due, in part, to bihemispheric effects since unilateral intracarotid amobarbital does not produce a completely pure unilateral cerebral dysfunction. However, acute unilateral hemispheric lesions can produce suppression of consciousness, even in the absence of direct brainstem involvement (Nichols, Mawad, Hilal, Mohr, Michelsen & Stein, 1985). Obviously, the attentional deficits produced by intracarotid amobarbital affect memory performance. However, left/right intracarotid amobarbital attentional deficits have been assumed to be equal, and thus the integrity of hippocampal structures becomes the critical factor. Although memory was significantly better following amobarbital injection on the nonfocus side as compared to the focus side in our study, the correct asymmetry was present in only 10 of the 18 patients with unilateral mesial temporal lobe focus. Our study demonstrated an inverse relation of perseverations to memory performance, but this effect does not explain the failure of the memory task to lateralize temporal lobe dysfunction. Since the reduction in memory associated with perseveration was primarily on the nonfocus side, this effect would have only enhanced the lateralization. However, amobarbital-induced alterations in attention or behavior that were not delineated by the simple task employed in our study could contribute to variance in memory performance.

Hemispheric Specializations

Differential roles have been suggested for the left and right cerebral hemispheres in attentional mechanisms. The left cerebral hemisphere appears to be specialized for attentional processes described as verbal, spanning, sequential, linear, or narrow, whereas the right cerebral hemisphere may predominate for visuospatial, scanning, holistic, gestalt or broad types of attention (Meador, Loring, Lee, Brooks, Thompson, Thompson & Heilman, 1988). The Wada test offers a unique opportunity to investigate the relative contributions of the left and right cerebral hemispheres to various cognitive functions. As it was originally designed to be used, the Wada test is employed clinically to assess lateralization of language, a primarily left cerebral function. In addition, several research studies have utilized the Wada test to assess lateralization of functions believed to be primarily mediated by the right cerebral hemisphere. We shall review three studies which have examined right cerebral specialization for various attentional mechanisms.

Medical College of Georgia Study on Tactile Attention. The major behavioral manifestations of the neglect syndrome include hemi-inattention, extinction to double simultaneous stimuli, hemispatial neglect and hemi-akinesia (Heilman, Valenstein & Watson, 1985). Although several components of neglect syndrome have been reported to occur more frequently following right cerebral lesions, a right cerebral predominance for directed tactile attention has not been statistically demonstrated. Findings of several studies that have demonstrated left/right asymmetries in tactile perception can be attributed simply to the task requirements involving the appreciation of spatial relationships (Carmon & Benton, 1969; DeRenzi, Faglioni & Scotti, 1970; Faglioni, Scotti & Spinnler, 1971).

In a study of tactile extinction employing complex stimuli and a verbal response in patients with unilateral cerebral pathology, Schwartz, Marchok, Kreinick & Flynn (1979) reported greater extinction of the left hand, but

Table 4-2 Demographics for MCG tactile inattention study.

Mean age = 35 years (Range=18-54)	
Sex: 9 men, 9 women	
Language dominance:	13 left
	4 mixed (L > R)
	1 mixed (R > L)
Handedness:	14 right
	2 left
	2 ambidextrous
Seizure focus:	7 left MTL
	4 right MTL
	2 bilateral MTL
	2 diffuse
	3 unknown
MTL = Mesial temporal lobe.	

(From Meador, Loring, Lee, Brooks, Thompson, Thompson & Heilman, 1988; reproduced with permission of *Neurology*)

attributed a slightly greater incidence of extinction in patients with right cerebral lesions to the selective exclusion of aphasics. However, when the data were divided by lesion site in each hemisphere, right parietal lesions produced significantly more extinction than left parietal lesions, even after taking into account the excluded aphasics. Schwartz, Marchok, Kreinick & Flynn (1979) proposed that this asymmetry was due to a relatively more widespread anatomical system in the right brain that made it more susceptible to random lesions than a constricted system on the left.

A specialized role for the right hemisphere has been postulated for attentional/intentional mechanisms directed toward external space (Heilman, Valenstein & Watson, 1985; Mesulam, 1981; Watson, Valenstein & Heilman, 1981). Since these mechanisms appear to be mediated by a distributed neuronal network including cortical and subcortical structures, it seems that the left/right asymmetry might exist on a functional basis rather than on differences in anatomic distribution. Therefore, as an alternative to the anatomic distribution hypothesis of Schwartz, Marchok, Kreinick & Flynn (1979), it has been proposed that a functional asymmetry for attentional/intentional systems exists between the left and right cerebral hemispheres. This functional hypothesis does not preclude that other anatomical asymmetries may underlie the functional asymmetry.

Therefore, we prospectively investigated the incidence of left/right tactile inattention during the Wada test (Meador, Loring, Lee, Brooks, Thompson, Thompson & Heilman, 1988). The Wada test offers the advantage of matching left/right brain lesions while studying the effects of transient cerebral

Table 4-3 Responses to left, right, and bilateral touch following each amobarbital injection

	Response				
Touch	L	R	B	NR	Total
Left Injection					
L	69	0	1	2	72
R	0	65	2	5	72
B	10	8	72	0	90
Right Injection					
L	49	8	0	15	72
R	1	67	3	1	72
B	1	44	45	0	90

(From Meador, Loring, Lee, Brooks, Thompson, Thompson & Heilman; reproduced with permission of *Neurology*)

dysfunction in each hemisphere of each individual. Further, the possible confounding influences of language impairment were controlled by employing a nonverbal task which was learned prior to onset of the cerebral lesion.

The subjects were 18 patients with intractable epilepsy who underwent the Wada procedure as part of their preoperative evaluation for epilepsy surgery. Demographic data are depicted in Table 4-2. Prior to Wada testing, subjects were instructed in and practiced a nonverbal test of tactile extinction. During the Wada procedure, subjects were tested in the supine position, and the examiner stood directly above the head of the angiography table. A stylized and distinctive figure was shown to the subject to denote the onset of the task. The subject would then close his eyes and place his hands on the sides of his chest (ipsilaterally). When touched only on the left hand, the subject was to respond by moving both hands to the left of his body. When touched only on the right hand, the subject was to move both hands to the right of the body. When touched simultaneously on both hands, the subject was to touch his chin with both hands. Thus, inattention to single stimuli contralateral to the amobarbital injection would result in no response. Further, extinction to simultaneous bilateral stimuli would produce movement toward the ipsilateral side. Even in the presence of contralateral hemiplegia, limb akinesia or hemispatial akinesia, extinction would be evident by movement of the ipsilateral arm toward ipsilateral hemispace rather than toward the chin.

Amobarbital dosages were determined on an individual basis, with incremental injections of sodium amobarbital (initial dosage = 50-75 mg, 25 mg increments) necessary to produce a contralateral flaccid hemiplegia. Mean left

Table 4-4 Comparison of tactile extinction studies

	Side of extinction	Left Lesions	Right Lesions
Schwartz et al. (1979)	L	22 (33%)	85 (99%)
	R	44 (67%)	1 (1%)
Meador et al. (1988)	L	8 (44%)	44 (98%)
	R	10 (56%)	1 (2%)

hemisphere dosage was 126 mg (SD=43); mean right hemisphere dosage was 127 mg (SD=38). Injections were administered by hand over a 4 second interval, and the order of injection was sequentially alternated. Following intracarotid amobarbital injections, patients initially underwent clinical evaluation which included tests of language, memory, spatial abilities, strength, and tone. The test for tactile inattention was then administered. All subjects were required to have test times postinjection within 100 seconds for the left and right injections. Mean starting and ending test times following injection were 363 and 445 seconds after left injection, and 367 and 443 seconds following right injection, respectively. Each subject underwent 13 trials. Four trials of unilateral touch to each hand and 5 trials of bilateral touch were administered in a randomized pattern.

More correct responses were given after left than right amobarbital injections (p<.0005). Responses were more likely to be correct for single contralateral than bilateral stimuli (p<.05). Tactile inattention occurred more frequently during right than left Wada (p<.0003). Double simultaneous stimulation was more sensitive to the inattention than single contralateral stimuli (p<.02). Extinction was significantly more frequent following right injection (p<.01). In a similar analysis for inattention to single contralateral stimuli, side was not significant, but in the 10 patients who exhibited inattention to single contralateral stimuli, 8 were after right Wada and only 2 were after left Wada. See Table 4-3 for data summary. These differences were not explained by any differences in left/right dosages, starting times, order of injection, or crossflow. Sex and seizure focus also had no effect on the results.

Conclusions. Thus, a left/right brain asymmetry for tactile attention was demonstrated with tactile inattention and extinction occurring more frequently during right Wada. In addition to extinction and inattention of single contralateral stimuli, another abnormal response pattern was apparent in the data (Table 3). Following right Wada injection, single left-sided touches elicited 8 right responses, whereas no left responses were elicited by single right-sided touches following left Wada injection. We postulated that the 8 responses contralateral to the side of stimulus represent a form of allesthesia

Table 4-5 Demographics for Boston visuospatial neglect study

Age Range = 18-50 years	
Sex: 16 men, 32 women	
Language dominance:	38 left
	7 right
	3 mixed
Handedness:	36 right
	12 left

(Data from Speirs, Schomer, Blume, Kleefield, O'Reilly, Weintraub, Osborne-Shaefer & Mesulam, 1990; reproduced with permission of *Neurology*)

(allochiria), which also exhibits a right cerebral predominance.

Several differences exist between our study and the previous investigation by Schwartz, Marchols, Kreinick & Flynn (1979). Yet, a comparison of the percent left and right extinction in the 2 studies is remarkably similar (Table 4-4). Schwartz, Marchols, Kreinick & Flynn (1979) hypothesized that somatosensory functions are more diffusely distributed in the right brain. However, we observed the same asymmetry for tactile extinction when the areas of left and right cerebral dysfunction were matched. Therefore, our hypothesis of a left/right functional asymmetry is more tenable.

Each cerebral hemisphere attends primarily to contralateral hemispace, and unilateral brain damage in either hemisphere can produce contralateral inattention. However, several investigations have suggested that the right brain may more easily attend toward either hemispace (Heilman & Van Den Abell, 1980; Reivich, Alavi & Gur, 1984; Weintraub & Mesulam, 1987). In our study, input from the intact left cerebral hemisphere did not compensate for the effects of right Wada inactivation on tactile attention as well as the right brain compensated for left Wada. Therefore, it appears that the right brain also has a greater capacity than the left to direct tactile attention toward either side of the body. The findings are consistent with prior studies that have demonstrated a right cerebral dominance for certain attentional/intentional mechanisms (Desmedt, 1977; Heilman & Van Den Abell, 1980; Reivich, Alavi & Gur, 1984; Weintraub & Mesulam, 1987). However, attention is not a unitary phenomenon, and various components of attention may be subserved preferentially by the left or right brain.

Boston Study on Visuospatial Neglect. Spiers, Schomer, Blume, Kleefield, O'Reilly, Weintraub, Osborne-Shaefer & Mesulam (1990) also employed the Wada test to investigate attentional mechanisms related to the neglect syndrome. In their study, they sought to demonstrate: 1) the presence of contralateral/ipsilateral visual neglect during the Wada test, 2) differential right/left cerebral effects, 3) relative interactions of cerebral dominance for

Table 4-6 Boston visuospatial neglect study: Mean contralateral and ipsilateral scanning performance at baseline and during right and left Wada tests for patients with left (L), mixed (M), and right (R) cerebral dominance for language. Scaled score of 4.0 denotes normal full scanning.

Hemispace Language	Contralateral				Ipsilateral		
	L	M	R		L	M	R
Baseline	4.0	4.0	4.0		4.0	4.0	4.0
Right Wada	1.1	2.0	1.7		2.8	3.2	3.1
Left Wada	3.8	3.3	3.4		4.0	3.7	3.9

(From Spiers, Schomer, Blume, Kleefield, O'Reilly, Weintraub, Osborne-Shaefer & Mesulam, 1990; reproduced with permission of *Neurology*)

language and handedness on these attentional functions, and 4) the degree and anatomical distribution of amobarbital suppression as measured by EEG.

The subjects were 48 consecutive patients who were candidates for epilepsy surgery. Demographics are presented in Table 4-5. Each patient underwent bilateral Wada tests as part of their preoperative evaluation. The left and right Wada tests were separated by 48 hours, and the order of administration was determined by site of targeted resection. A single bolus intracarotid injection of 175 mg amobarbital was given for each procedure. The experimental tasks were interspersed throughout the standardized Wada examination. The primary dependent measure was a visuospatial task modified from the Random Letter Cancellation Test (Weintraub & Mesulam, 1985, 1988). The subject was required to scan a page of scattered letters to find the target letter "A" and point to every "A" on the sheet with the hand ipsilateral to the injected cerebral hemisphere. The patient's performance was recorded on audiotape dictating the starting point, direction, and spatial course adopted for scanning. Performance for scanning of contralateral and ipsilateral space was scored on a 4 point scale which assessed scanning of quadrants, midline, and edges of the stimulus sheet. EEG was also recorded simultaneously and later rated by a blinded observer for slowing on a 5 point scale. Subjects were tested prior to the Wada test and following amobarbital when they showed their first appropriate response to the stimulus card. The first appropriate response occurred between 3-12 minutes post left Wada injection and usually within 5 minutes post right Wada injection. If subjects exhibited visual neglect on the scanning task, they were retested as the effects of amobarbital resolved.

The visuospatial scanning task was performed without error by all subjects at baseline. In dextral subjects with left cerebral language dominance, the right hemisphere injection resulted in significant (p <.001) disruption of contralateral and ipsilateral scanning, whereas left hemisphere injection produced only minor, nonsignificant, disruption of contralateral scanning and had no effect on ipsilateral scanning. Further, the effect of the right cerebral injection was greater for the contralateral than the ipsilateral space (p<.01), although both were significantly depressed compared with either the control or left hemisphere injection conditions. Left-handed subjects with atypical dominance differed from right-handed subjects with typical dominance in that the right hemisphere injection significantly disrupted only contralateral (p<.001), but not ipsilateral, scanning. As in the subjects with typical dominance, left injection had no effect on scanning performance in either the contralateral or ipsilateral space. Patients with right-hemisphere language dominance also showed contralateral scanning disruption only during right-hemisphere suppression (p<.01). See Table 4-6 for a summary of the data. In the great majority of our subjects, the left neglect observed during right Wada appeared to occur in the absence of hemianopsia. Analysis of the EEG revealed significantly more slowing in the right prefrontal and frontal regions of subjects with neglect compared to those without neglect (p<.001). No significant effects were found in other brain regions. A similar phenomenon was noted for intrasubject EEG during resolution of the amobarbital-induced neglect.

Conclusions. Although the test times after left and right Wada injections are not specifically matched, the results of this study are consistent with a right cerebral dominance for control of directed attention in extrapersonal space. The scanning impairments in both contralateral and ipsilateral space during right Wada support the notion that the right cerebral hemisphere can direct attention toward both contralateral and ipsilateral hemispace, whereas the left cerebral hemisphere directs attention only toward contralateral hemispace (Heilman & Valenstein, 1979; Mesulam, 1981; Weintraub & Mesulam, 1987). The intersubject and intrasubject observations that amobarbital-induced EEG slowing in the frontal, but not parietal or temporal lobes, was related to the presence of neglect suggests that the visuospatial scanning deficit is related to a directional hypokinesia and to a failure of frontal mechanisms controlling exploratory-motor aspects of attention directed at the external milieu (Heilman, Bowers, Coslett, Whelan & Watson, 1985; Spiers, Schomer, Blume, Kleefield, O'Reilly, Weintraub, Osborne-Shaefer & Mesulam, 1990). Finally, Spiers, Schomer, Blume, Kleefield, O'Reilly, Weintraub, Osborne-Shaefer & Mesulam (1990) concluded that the lateralized control for attention to extrapersonal space is largely independent of the lateralization of language and handedness. However, the means on the scanning task (see Table 4-6) and some of the statistical analyses suggest that atypical cerebral dominance for language and handedness does affect lateralization of these attentional mechanisms to some degree. In addition, the next study demonstrates that atypical dominance can affect the expression of supranuclear attentional/intentional mechanisms controlling eye gaze.

Medical College of Georgia Study on Eye Gaze. The relationship between unilateral cerebral lesions and conjugate lateral gaze has been well demonstrated through a variety of investigations employing clinicopathological

Table 4-7 Subject demographics for MCG eye gaze study

Mean Age = 30 years (Range=13-61)		
Sex: 49 males, 41 females		
Handedness:	82 right	
	7 left	
	1 ambidextrous	
Language dominance:	75 left	
	15 mixed	
Seizure focus:	33 left	
	32 right	
	9 bilateral	
	7 diffuse	
	9 unknown	

(From Meador, Loring, Lee, Brooks, Nichols, Thompson, Thompson & Heilman, 1989; reproduced with permission from *Brain*)

correlations and stimulation and/or lesioning techniques (Bender, 1980; Bizzi, Kalil & Tagliasco, 1971). Voluntary activation or electrical stimulation of one cerebral hemisphere produces turning of the eyes and head toward contralateral hemispace. In contrast, a lesion of 1 cerebral hemisphere typically produces eye and head turning toward ipsilateral hemispace. Since the effects of the 2 cerebral hemispheres are balanced when the eyes are in primary position, the ipsilateral conjugate gaze deviation associated with unilateral cerebral hemispheric lesions reflects a deficit in the ipsilateral supranuclear gaze mechanisms. This phenomenon was initially reported by Andral (1834) and first investigated systematically by Prevost (1868). The clinicopathologic study of 51 patients by Prevost remains the largest autopsy-verified series, and he also confirmed his clinical observations with animal experiments. Despite numerous subsequent investigations, a left/right cerebral asymmetry for gaze mechanisms was not noted until 1982 when De Renzi, Faglioni & Scotti reported that gaze paresis was more frequent (72/120), severe, and persistent in patients with right cerebral lesions. They postulated that the left/right gaze differences were due to an asymmetric anatomical distribution of the oculomotor system in the 2 hemispheres, with the system represented more diffusely on the left and more focally on the right. However, these conclusions were based on clinical data, and the majority of cases had neither radiological nor pathological verification of lesion distribution. Mohr, Rubinstein, Kase, Price, Wolfe, Nichols & Tatemichi (1984) also reported a right cerebral predominance for gaze paresis in 531 cases of hemisphere stroke, but an asymmetric left/right distribution of deep lesions precluded firm conclusions concerning left/right hemispheric mechanisms.

In view of the specialized role for the right hemisphere in attentional and intentional mechanisms directed toward external space (Heilman &

Table 4-8 Eye gaze after Wada injection (n = 90)

Injection	Left	Midline	Right
Left	29	58	3
Right	1	35	54

(From Meador, Loring, Lee, Brooks, Nichols, Thompson, Thompson & Heilman, 1989; reproduced with permission from *Brain*)

Valenstein, 1979; Heilman, Valenstein & Watson, 1985; Mesulam, 1981), we presented an alternative hypothesis to the anatomical postulate of De Renzi. We proposed that a functional asymmetry for attentional/intentional systems exists such that the right cerebral hemisphere may direct the eyes toward either hemispace, while the left cerebral hemisphere is limited in its capacity to direct the eyes ipsilaterally. Therefore, we investigated prospectively the left/right incidence of gaze paresis during the Wada test (Meador, Loring, Lee, Brooks, Nichols, Thompson, Thompson & Heilman, 1989).

Ninety patients with intractable epilepsy underwent the amobarbital procedure as part of their preoperative evaluation for epilepsy surgery. Demographic data are depicted in Table 4-7. Amobarbital dosages were determined on an individual basis. All patients received incremental injections of sodium amobarbital (initial dosage = 50-75 mg, 25 mg increments) necessary to produce a contralateral flaccid hemiplegia. Mean left hemisphere dosage was 106 mg (SD = 42.1); mean right hemisphere dosage was 101 mg (SD = 36.6). Injections were administered by hand over a 4 second interval, and the order of injection was alternated sequentially.

Patients were tested in the supine position. The primary examiner stood directly above the head of the angiography table. Assistants stood to the left and right of the table. At the start of the test, the patient held his arms straight up with the hands supinated and fingers extended. He/she was instructed to look straight ahead and count repeatedly from 1 to 20. The amobarbital injection was then administered. At the onset of flaccid hemiplegia, patients were instructed to stop counting if speech arrest had not occurred. Observations were then made for the presence of spontaneous, tonic (i.e., >2 sec) eye deviations from the midline. Examiners to the left and right of the angiography table called for the patient to look toward the contralateral and then the ipsilateral side. (Patients were instructed prior to the injection as to the procedure of looking to the left and right on command). Cerebral dominance for propositional language was determined based on speech arrest, comprehension, repetition, and naming. All sessions were videotaped. The second half of the series (i.e., 45 patients) was also rated (0-3) for severity of eye deviation. The severity ratings were made according to the following scale: 0 = no deviation, 1 = transient eye deviation of less than 10 seconds (without

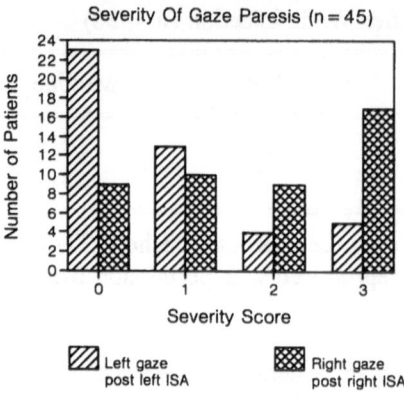

Figure 4-1. Gaze paresis ratings following Wada injections (none = 0, mild = 1, moderate = 2, severe = 3). Note greater severity of gaze deviation following right Wada test (From Meador, Loring, Lee, Brooks, Nichols, Thompson, Thompson & Heilman, 1989; reproduced with permission from *Brain*).

head deviation), 2 = eye deviation of 10-30 seconds duration ($\pm$ head deviation), and 3 = substained eye deviation greater than 30 seconds (usually with associated head deviation). Final ratings required the agreement of the 3 observers. Any disagreements were reviewed on videotape to reach a unanimous rating.

When gaze deviation occurred, the eyes moved conjugately toward the ipsilateral side in almost every case. The deviation was tonic and associated with the development of hemiplegia. Following left Wada injection, patients usually did not respond to command and exhibited few spontaneous saccades immediately post-injection. Following right Wada injection, patients were able to perform pursuit and saccade maneuvers on command but were frequently unable to direct gaze volitionally past midline. Since aphasia interfered with testing of eye movements to command following left Wada injection, we chose to limit our analysis to the left/right Wada difference in spontaneous tonic eye deviation. The raw data are presented in Figure 4-1 and Tables 4-8 & 4-9. After left Wada, 32% of the patients exhibited ipsilateral gaze deviation and 100% exhibited some language dysfunction. After right Wada, 60% had gaze deviation and 17% had language deficits. In the group with greatest cerebral lateralization (i.e., dextrals with left language dominance), 26% exhibited ipsilateral eye deviation after left Wada and 61% after right Wada. In the group with mixed cerebral dominance, 55% exhibited ipsilateral gaze deviation after left Wada and 55% after right Wada.

Gaze paresis was more frequent after the right injection (p<.0002). This effect was even more marked (p< .00006) when the analysis was limited to dextrals with left language dominance. Cerebral lateralization significantly

Table 4-9 Wada eye gaze deviation related to cerebral dominance

Eye gaze in dextrals with left language dominance (n = 70)			
Injection	Left	Midline	Right
Left	18	49	3
Right	1	26	43
Eye gaze in patients with mixed dominance (n = 20)			
Injection	Left	Midline	Right
Left	11	9	0
Right	0	9	11

(From Meador, Loring, Lee, Brooks, Nichols, Thompson, Thompson & Heilman, 1989; reproduced with permission from *Brain*)

affected Wada eye gaze results (p< .05). In fact, subjects with mixed cerebral dominance for handedness and/or language exhibited no difference in the incidence of eye gaze deviation following left and right Wada injections. Overall, gaze paresis was significantly more severe after right versus left Wada injections (p<.001). As reflected in the rating scores, the duration of deviation was longer after right Wada. In some cases (almost always after right Wada), the duration of the eye deviation was prolonged for several minutes (e.g., the longest duration was greater than 9 minutes). In these patients, the deviation required presentation of test stimuli in the extreme right visual field. The effects were not explained by left/right differences in amobarbital dosages, order of injection, crossflow, or seizure focus.

Conclusions. A left/right brain asymmetry in gaze mechanisms was demonstrated with the incidence and severity of gaze paresis being greater for right cerebral dysfunction. The findings support a hypothesis of a functional asymmetry in gaze mechanisms between the left and right brain. Input from the intact left cerebral hemisphere did not overcome the effects of right Wada on gaze as well as the intact right brain compensated during left Wada. Therefore, it appears that the right brain has a greater capacity to direct the eyes toward either hemispace than the left brain which appears limited in its ability to direct the eyes ipsilaterally. Although eye gaze is largely a motor phenomenon and reflects mainly intentional mechanisms, attentional and intentional mechanisms function interactively (Goldberg & Segraves, 1987). Therefore, we postulated that the gaze asymmetry is due to right cerebral dominance for attentional/intentional mechanisms directed at external space. Based on scalp recorded presaccadic potentials, Moster & Goldberg (1990) offered an alternative explanation suggesting that dextrals with left language dominance have an innate right gaze preference which produces greater rightward eye deviation during right Wada than leftward deviation during left Wada. Similar turning behaviors have been related to neurochemical asymmetries in animals (Zimmerberg, Glick & Jerussi, 1974; Pycock, 1980). However, the explanation of Moster & Goldberg (1990) is not at odds with our hypotheses. If the direction of motor attention (and thus eye gaze) is predominantly toward contralateral hemispace for the left brain but more equally balanced across hemispace for the right brain, then subjects with typical cerebral dominance would indeed exhibit a right gaze preference.

The difference in gaze paresis between the group with greatest cerebral lateralization (i.e., dextrals with left cerebral language) and those with mixed cerebral dominance is noteworthy. The cerebral asymmetry for gaze paresis was present only for dextrals with left language dominance suggesting a relationship of cerebral lateralization to gaze mechanisms. The left hemisphere is specialized for language and praxis, while the right hemisphere is specialized for complex visuospatial perceptual tasks, some aspects of emotional behavior, and the spatial distribution of attention and intention (Heilman & Valenstein, 1985). Mesulam (1985) has suggested 2 hypotheses concerning the evolutionary development of the right hemisphere specializations. According to the first, the specialized non-linguistic functions of the right brain were initially equally distributed between the 2 hemispheres, but during the subsequent development of language, the left brain's non-linguistic abilities were diminished. Thus, the right brain specializations appeared secondary to a reduced capacity of the left brain. Since the left brain retains some of its prior functions, this hypothesis predicts that left brain lesions would disrupt non-linguistic functions to a greater degree than equivalent lesions of the right brain would impair language. According to the alternate hypothesis, the right brain specializations represent emergent functions which evolved simultaneously with left cerebral language. This hypothesis predicts that left brain lesions would disrupt non-linguistic functions to approximately the same degree that right brain lesions impair language. In our study, left Wada produced gaze deviations more frequently than right Wada produced language deficits, and this effect was more pronounced in the group with greatest cerebral lateralization (i.e., dextrals with left language dominance). Further, the occurrence of gaze deviation after either left or right Wada in subjects with mixed cerebral dominance was comparable to the occurrence after right but not left Wada in dextral subjects with left cerebral language dominance. The results suggest that the role of the left brain in directed spatial attention/intention diminished as it evolved language specialization. Whether right brain dominance for more complex non-linguistic functions developed by a similar process or by emergent evolution is uncertain.

5
Emotion

The focus of biological theories of emotion may be viewed as having traveled from the lower-peripheral areas (e.g., James-Lange visceral theory) to the higher-central regions involving the cortex. Despite having traveled up the neuraxis, most biological theories of emotion continue to rest on the foundations first suggested by William James in 1884. According to James, there is an intimate relationship between visceral-autonomic functions and emotional expression and experience. Although this notion has continued acceptance, this relationship was criticized by Cannon (1927) who observed that the same visceral-autonomic changes occur in very different emotions and may even occur in the absence of emotion. Visceral-autonomic changes accompanying emotion provide a general, final common pathway of emotional expression (Pribram & Melges, 1969).

Stimulation or ablation of various regions of cortex which are not part of the limbic system can result in visceral or emotional changes. Thus, emotional theories have been challenged at every step, since those autonomic and emotional functions that have been attributed to a specific region are not restricted to that region alone. As the study of emotions has progressed from peripheral to central, the participation of the highest levels of the neuraxis in emotional expression and experience has been increasingly acknowledged. Heilman, Watson & Bowers (1983) observed:

> an emotion thus depends on varied anatomic structures, including: cortical systems for producing the appropriate cognitive set, limbic structures for activating the brainstem and thalamic activating centers and for controlling hypothalamic output, the hypothalamus for regulating endocrine and autonomic responses, and the brainstem and thalamic activating systems for producing cortical arousal (p. 58).

Emotional Perception and Expression

The emotional disorders seen in neurologic patients with unilateral lesions are generally consistent with the results of Wada studies. There are a large number of clinical and experimental reports dealing with the receptive and expressive emotional capacities of medical patients and normals. We briefly discuss this line of research below. Comprehensive reviews of this literature may be found in Tucker (1981), Heilman, Bowers & Valenstein (1985), and Silberman & Weingartner (1986).

Among patients with unilateral damage to a single hemisphere of the brain, Hécaen (1962) found a greater incidence of catastrophic reactions in left-sided lesions (55 of 206 cases, or 27%) as compared with right-sided lesions (20 of 154 cases, or 13%). In contrast, Hécaen (1962) reported indifference to failures (anosodiaphoria) was observed in 34 of 206 left-sided lesion cases (16%) while indifference reactions were seen in 51 of 154 right-sided cases (33%). These findings were confirmed by Gainotti (1969) who found that depressive-catastrophic reactions were most common among patients with damage to the left hemisphere (30 of 80 lefts, or 38% vs. 7 of 70 rights, or 10%) and indifference-euphoric reactions more often in cases with right hemisphere disease (23 of 70 [33%] versus 9 of 80 [11%] lefts). Catastrophic reactions in these investigations included increasing signs of anxiety or sudden bursts of tears. Indifference reactions involved indifference toward neuropsychological test failures, ignoring the obvious consequences of their disabilities, or treating their failures and disabilities with cheerfulness and joking.

Several lines of investigation have suggested that the right hemisphere is dominant for both the reception and expression of emotions (Mesulam, 1985). Patients with damage to the right hemisphere show less facial expressiveness and have more difficulty in interpreting the emotions of others through facial expressions and vocal prosody than do patients with left hemisphere lesions (Tucker, Watson & Heilman, 1977; Buck & Duffy, 1980). In neurologically intact individuals, the perception of emotions seems to be more accurate, and they display more expression with activation of the right hemisphere (Campbell, 1978; Ley & Bryden, 1979).

These data appear to be related to the specialized perceptual capabilities, holistic, non-linear processing style, and relative dominance for attention and arousal of the right hemisphere. It is difficult to integrate the important role of the right hemisphere in processing emotional information with hemispheric specialization for specific emotions (e.g., depression) or for certain intrinsic emotional tones (e.g., positive or negative emotions). Further information is necessary before this integration can be made accurately.

Major Empirical Investigations

Early Italian Investigations. Terzian & Cecotto (1959) were the first to report that depressive-catastrophic reactions were seen after amobarbital injection into the left hemisphere, while euphoric-maniacal or indifference reactions were more common following right hemisphere amobarbital injections. Soon after, other Italian investigators reported similar findings (Alema & Donni, 1960; Perria, Rosadini & Rossi, 1961; Terzian, 1964; Alema & Rosadini, 1964; Rossi & Rosadini, 1967).

In the first English language publication concerning lateralized emotional reactions, Perria, Rosadini & Rossi (1961) reported depressive reactions following anaesthetization of the speech dominant hemisphere, with euphoric type responses more frequently seen with nondominant injections. After dominant hemisphere injection, "the patient cries and says that he will never recover, that his family will go to ruin," and during nondominant injections "he is optimistic about his future, makes jokes, and often breaks into laughter" (p. 74). These emotional reactions were observed 4-6 minutes following injection and lasted between 1 and 10 minutes. The emotional responses were typically accompanied by a return to electroencephalographic fast activity indicating late amobarbital EEG effects. The emotional reactions were also observed when all EEG abnormalities had disappeared. Further, no emotional reaction was observed in approximately one-third of dominant hemisphere injections and one-half nondominant injections. However, of the 30 patients included in the study only 11 received injections to both hemispheres and dosage of amobarbital varied widely (from 2.5 to 100 mg).

When Terzian (1964) injected 100 mg of amobarbital into the left carotid artery, "towards the end of the aphasic episode...the patient, especially when spoken to, despairs and expresses a sense of guilt, of nothingness, of indignity, and worries about his own future or that of his relatives, without referring to the language disturbances overcome and to the hemiplegia just resolved and ignored" (p. 235). Injection into the right carotid artery was reported to cause a euphoric reaction which "in some cases may reach the intensity of a maniacal reaction." In the euphoric emotional response, "the patient appears without apprehension, smiles and laughs and both with mimicry and words expresses considerable liveliness and sense of well being" (p. 235). Although the number of reactions and number of patients examined were not reported, the 2 emotional reactions did not occur in every patient. Depressive-catastrophic reactions were only observed when carotid injection produced language disturbances, and euphoric-maniacal reactions occurred exclusively with injection to the side where no speech alterations were seen. As in previous investigations, not all patients received amobarbital injections to both hemispheres.

University of Genoa Study. Rossi & Rosadini (1967) were the first investigators to report specific numerical results of emotional reactions following amobarbital injections. Of the 126 patients examined, 49 received injections to both hemispheres, and 77 injections were made on a single side only. Amobarbital doses ranged from 100-200 mg (in 5% solution) administered over a 4-5 second period into the first portion of the internal carotid artery in most cases, and into the common carotid in others. Emotional reactions occurred in 73 of the 175 injections (42%). After excluding patients with suboptimal examinations (abnormal vasculature, bilateral drug distribution, drug dose too low to produce hemiparesis), emotional reactions occurred in 53.7% (73/136) of injections. The appearance of the emotional reactions in relation to the time of injection was "variable," but in the majority of cases the reactions were seen later in the period of drug effect. "Depressive-catastrophic" emotional reactions were observed in 25 injections, 18 after left and 7 after right hemisphere injection. Reactions of the "euphoric-maniacal" type occurred in 44 injections, 11 of them left and 33 right hemisphere injections. In 4 injections, all into the right hemisphere, there were manifestations of both depression and euphoria. In 5 patients, euphoria was elicited following both left and right hemisphere injections. These results are summarized in Table 5-1. Notice that the data in Table 5-1 refer to the number of injections, not to the number of individual patients.

Rossi & Rosadini concluded there was a definite prevalence of depressive reactions following left, and of euphoric reactions following right, hemisphere barbiturization. Similar emotional specialization was found when they examined the data by handedness and cerebral speech dominance. Unfortunately, Rossi & Rosadini did not state the total number of left and right side injections (77 patients received unilateral injections, side unspecified), but only the number of emotional reactions associated with left and right injections. Thus, it is impossible to determine the relative prevalence of the various emotional categories. For example, there were 18 depressive-catastrophic reactions after left injection and 7 after right, which suggests depressive responses are more common following left hemisphere inactivation. If, however, there were more left than right sided injections, this apparent difference could disappear.

Montreal Neurologic Institute Study. At the conclusion of the Rossi & Rosadini (1967) paper, Milner summarized the results of a similar study of emotional responses following amobarbital injection which she and Branch conducted at the Montreal Neurologic Institute (Milner & Branch, 1967). These data are presented in Table 5-2. Mood ratings were made on a 5-point scale following unilateral injections in 104 patients. All patients received injections of 200 mg (in 10 percent solution) into the common carotid artery on each side. Of the 104 patients, 40 (39%) showed identical mood ratings after both left and right injections. However, patients displaying no reactions were not differentiated from those with similar reactions following both injections. Of the remaining 64 patients, 39 (37%) seemed more euphoric after left than right injection and 25 (24%) appeared to be more euphoric with right rather than left injection. Only 5 depressive reactions were observed, 3 following dominant and 2 after nondominant injections.

Table 5-1 Emotional reactions following amobarbital injection

	Left Injection	Right Injection	Total
Depression	18 (62%)	7 (16%)	25
Euphoria	11 (38%)	33 (75%)	44
Both	0 (0%)	4 (9%)	4
Total	29 (100%)	44 (100%)	

(From Rossi & Rosadini, 1967; adapted with permission of Grune & Stratton)

Consistent with her data, Milner concluded there was no association of depressive reactions with left, and of euphoria or elation with right, hemisphere amobarbital injections.

Other investigations have observed no changes in emotional expression suggestive of hemispheric specialization following intracarotid injections of amobarbital. Werman, Christoff & Anderson (1959) reported no mental changes in 16 patients after amobarbital injections of 25-75 mg. Tengesdal (1963) injected 25-100 mg of amobarbital into the carotid artery of an unknown number of patients and although "emotional changes have been tested for, we have so far not been able to confirm the results of Perria, Rosadini & Rossi (1961) who described a depressive emotional state when injecting the dominant hemisphere, and a euphoric reaction on the non dominant side" (p. 332). After injecting 125 mg of amobarbital into the carotid arteries of 12 epilepsy surgery candidates, Fedio & Weinberg (1971) failed to observe the lateralized depressive and euphoric emotional responses.

In contrast, Deglin & Nikolaenko (1975) obtained results strikingly similar to Rossi & Rosadini (1967) when they judged changes in affect and mood following unilateral electroconvulsive shock therapy in 40 psychiatric patients (see Table 5-3). Similarly, in a more recent Wada study, Silfvenius, Blom, Nilsson & Christianson (1984) reported that drug injection (150-175 mg, 10 % solution) resulted in emotional reactions in 10 of their 18 (56%) patients. Weeping, laughing, and unrest were observed in 8 cases after injection to the speech dominant hemisphere, and after right hemisphere injection, unrest was seen in 2 cases.

There are then essentially an equal number of studies with diametrically opposed results. Some of the factors to be considered in explaining these different conclusions include amobarbital dose, duration and pattern of crossflow of the drug to the opposite hemisphere, type of neurologic disease of the patient under study, criteria used to determine what constitutes an emotional response, definition of each emotional reaction, and perhaps the premorbid personality characteristics of the patient under study.

Table 5-2 Emotional reactions following amobarbital injection

	Left Injection	Right Injection	Total
Depression	3 (7%)	2 (7%)	5
Euphoria	39 (93%)	25 (93%)	64
Total	42 (100%)	27 (100%)	

(Adapted from Milner & Branch, 1967)

Medical College of Georgia Study. Because controversy existed concerning alternations in emotional expression following intracarotid amobarbital injections, we examined the number and type of emotional reactions during the Wada procedure in 55 (31 male, 24 female) epilepsy surgery patients. Forty-four of these patients were the subject of an earlier investigation (Lee, Loring, Meador & Brooks, 1990). All patients received incremental injections of amobarbital (up to 250 mg in 5 ml solution) necessary to produce a dense, contralateral hemiplegia. Mean left hemisphere dosage was 133 mg and mean right hemisphere dosage was 123 mg. Depression was defined by sustained crying, tearfulness, or sobbing with sad affect and was often accompanied by depressive verbalizations with themes of discouragement or worthlessness. Euphoria was rated if there were sustained periods of laughter or an elated and expansive mood as evidenced by giggling, cheerfulness, or excessive talking, with speed often being rapid, loud and at times pressured. To be classified as a definite change, emotional reactions were required to persist for at least 30 seconds.

These 55 patients showed 28 emotional reactions, 13 reactions after left and 15 following right intracarotid injections. The majority of patients showed no

Table 5-3 Emotional reactions following unilateral ECT

	Left ECT	Right ECT	Total
Depression	33 (94%)	5 (10%)	38
Euphoria	2 (6%)	47 (90%)	49
Total	35 (100%)	52 (100%)	
No Reaction	36	34	70

(From Deglin & Nikolaenko, 1975)

Table 5-4   Emotional reactions following amobarbital injections

	Left Injection		Right Injection	
	Depression	Euphoria	Depression	Euphoria
Rossi & Rosadini	18 (62%)	11 (38%)	7 (17%)	33 (83%)
Milner & Branch	3 (7%)	39 (93%)	2 (7%)	25 (95%)
Lee et al.	8 (62%)	5 (38%)	0 (0%)	15 (100%)

emotional reaction after amobarbital injection to either hemisphere. Among the 28 reactions observed, left hemisphere injection produced depression-crying in 8 and euphoria-laughter in 5 patients. Right hemisphere injection produced 15 cases of euphoria-laughter, and no patient became depressed or cried. Depression-crying was seen only following left hemisphere injections, while euphoria-laughter was more common after right, rather than left, hemisphere injections.

These results are compared with the studies of Rossi & Rosadini (1967) and Milner & Branch (1967) in Table 5-4. Examination of Table 5-4 reveals an identical incidence of affect change following left hemisphere injection between the Rossi & Rosadini sample and our patients, while the Milner & Branch sample differs considerably with regard to the number of both depressive and euphoric reactions. In contrast, all 3 samples displayed similar frequencies of emotional reactions after right hemisphere injection.

Causes of Amobarbital Emotional Reactions

Dosage and Degree of Drug Crossflow. Amobarbital dose does not appear to be an important factor in causing emotional reactions. Rossi & Rosadini (1967) reported observing lateralized emotional reactions with an amobarbital dose of 300 mg in 1 patient and with doses too low to create even mild impairments of sensorimotor and language functions in others. Similarly, the duration and pattern of crossflow of the drug to the opposite hemisphere or to the ipsilateral posterior cerebral artery cannot explain the lateralized pattern of emotional response. In our study (Lee, Loring, Meador & Brooks, 1990), there were no left/right differences in the presence or persistence of angiographic crossflow. Since the degree of crossflow was essentially the same for left and right amobarbital injections, the lateralized emotional reactions were not the result of generalized, diffuse effects of amobarbital.

There is also evidence that crossflow observed with mechanical pressurized injections used during conventional angiography is not reflective of crossflow following the slower hand-injected amobarbital. Jeffery, Monsein, Szabo, Hart, Fisher, Lesser, Debrun, Gordon, Wagner & Camargo (1991) observed contralateral flow in 68% of patients during conventional angiography while crossflow, as measured by SPECT, was not seen in any patient undergoing amobarbital hand injection.

Type of Disease. The type of neurologic disease has been posited as a possible explanation for the different results in the Rossi & Rosadini (1967) and Milner & Branch (1967) studies. Only 49 of the 126 patients in Rossi & Rosadini's sample were epileptic. The remaining patients included cases of vascular disease, extrapyramidal disease, tumor, and psychiatric illness. Depending upon the location of lesion and diagnosis, many of these patients could be considered already at-risk for the development of some form of emotional reaction. Although not explicitly stated, it may be assumed that most if not all of the Milner & Branch patients were epilepsy surgery candidates and all 55 cases in our study were epilepsy patients. Since our results are so similar to those of Rossi & Rosadini's, despite the differences in diseases comprising the samples, the type of neurologic disease is not likely the crucial reason for the difference in reported lateralized emotional reactions.

Premorbid Personality. Milner & Branch (1967) noted that characteristically jovial patients tended to become euphoric, and placid individuals showed little affect change following intracarotid amobarbital injection. Because similar reactions following injections to both hemispheres in the same individual were observed, Milner suggested that premorbid personality characteristics or temperament was an important factor in determining which patients will show emotional reactions and may also influence the type of reaction displayed. The Rossi & Rosadini (1967) and Lee, Loring, Smith & Flanigin (1990) results tend to downplay the role of premorbid personality features since depressive and euphoric reactions may occur in the same patient following injections to opposite hemispheres. The presumed role of premorbid personality in the genesis of emotional reactions is not restricted to amobarbital injection. Brain damage of any type may cause an accentuation of premorbid personality traits, such that characteristically outgoing individuals become euphoric or the more reserved and shy become depressed. Others have emphasized that the damaged region of the brain is more important than premorbid temperament in the creation of mood changes after brain injury or disease.

To explore the role of premorbid personality factors in emotional reaction during the Wada test, we examined the Minnesota Multiphasic Personality Inventories (MMPI) of 26 (10 male, 16 female) epilepsy surgery candidates who showed either a euphoric/laughter or depressive/crying reaction after unilateral amobarbital injection. If the accentuation of premorbid personality traits hypothesis is correct, the MMPI Hypomania Scale should be higher among patients who showed euphoric reactions and the MMPI Depression Scale should be higher among those with depressive reactions. Alternatively, if laterality of lesion were more important that premorbid personality factors,

Table 5-5 Mean (SD) MMPI T-Scores by type of emotional reaction following amobarbital injection

	Euphoria/Laughter	Depression/Cyring
Hypomania	67.2 (12.1)	61.5 (8.3)
Depression	63.1 (11.3)	60.2 (11.5)

the type of emotional reaction should be associated with inactivation of a specific hemisphere.

Epilepsy surgery candidates who showed either a euphoric/laughter or depressive/crying response, as defined above in our previous study, during the Wada test and who had completed a valid MMPI within 2 weeks of the Wada evaluation were selected for review. Mean amobarbital dose on the side of injection eliciting the emotional reaction was 117 mg. Means (and standard deviations) of MMPI T-scores by type of emotional reaction are given in Table 5-5.

There were no statistically significant differences between euphoric/laughter and depression/crying reaction patients on MMPI Scale 9 or on MMPI Scale 2. With regard to the association between hemispheric inactivation and type of emotional reaction, 14/18 (78%) of euphoric/laughter reactions followed right hemisphere amobarbital injection and 8/8 (100%) of depressive/crying responses followed left hemisphere injection.

These results suggest that premorbid personality factors, as measured by the MMPI, are not the primary contributor to lateralized emotional reactions following unilateral cerebral lesions. Depressive/crying reactions were clearly related to left hemisphere inactivation and euphoria/laughter were more common after right hemisphere inactivation. Thus, the location of cerebral lesion appears to be more important in the genesis of emotional reactions after lateralized brain damage than are premorbid personality traits.

Hemispheric Specialization for Emotion

Converging lines of evidence from lesion, amobarbital, and electro-convulsive therapy studies indicate that euphoria more frequently occurs after right hemisphere damage and depression more frequently following left hemisphere damage. The interpretation given to lateralization for positive and negative emotions is, however, frequently debated (e.g., Kurthen, Linke, Reuter, Hufnagel & Elger, 1991).

There are 3 primary explanations for hemispheric emotional specialization. Some have proposed that emotional changes after brain damage are secondary psychological reactions to physical or cognitive disability. The other 2 hypotheses attribute emotional reactions to the direct effects of brain damage, but differ as to which side of the brain is causing the excessive emotional expression. Some have speculated that emotional changes are caused by the altered functioning of the lesioned hemisphere or ipsilateral subcortical structures. Others have hypothesized that the emotion expressed after damage to 1 cerebral hemisphere reflects the intrinsic emotion of the opposite, intact hemisphere. That is, when 1 hemisphere is damaged or inactivated, it cannot contribute to behavior. In fact, its deactivation may result in a loss of inhibition over the opposite hemisphere causing the unrestrained emotional tendencies of the opposite hemisphere to be freely expressed.

The mood changes seen after cerebral insult may not necessarily be caused directly by disrupting the neural mechanisms underlying emotion, but rather, may be secondary psychological reactions to physical and cognitive disabilities (Hécaen, 1962; Gainotti, 1972; Williams, Little & Klein, 1986). Thus, crying and depression after dominant hemisphere injection might be seen as the patient's emotional response to hemiplegia and aphasia. This notion has not been well supported by research. In investigations of depression following stroke, the location of lesion was more strongly related to the mood disorder than were impairments in social functioning (Robinson & Price, 1982; Robinson, Starr, Kubos & Price, 1983), severity of aphasia (Robinson, Kubos, Starr, Rao & Price, 1984), degree of sensorimotor deficit, and severity of cognitive impairment (Robinson, Starr, Lipsey, Rao & Price, 1985; Gasparrini, Satz, Heilman & Coolidge, 1978). Further, in our experience, the majority of patients exhibiting emotional reactions after amobarbital injection do not notice or recall their hemiplegia or aphasia, yet many nevertheless continue to express some form of emotional response (Lee, Loring, Smith & Flanigin, 1990). Finally, this explanation may intuitively make sense to explain the sadness seen after left hemisphere amobarbital injection, but it does not account for euphoria after right injection. Why would an individual feel elated and happy because he was left hemiparetic and had left neglect and a visuospatial disorder? Taken together, these factors indicate that affect and mood changes after brain insult are manifestations of the underlying brain damage and are not secondary psychological reactions to the consequences of the damage.

Emotional changes following brain damage may also be caused by the altered functioning of the lesioned hemisphere or ipsilateral subcortical influences. Tucker (1981), and Tucker & Frederick (1989) have proposed that the primary effect of a unilateral lesion is to disinhibit the more primitive emotional functions of the lesioned hemisphere or its subcortical connections. The main support for the ipsilateral release hypothesis comes from Rossi & Rosadini's (1967) anecdotal observation that amobarbital emotional reactions only occurred after the motor, sensory, speech, and EEG changes caused by the drug have disappeared. This was estimated to be 4-6 minutes after amobarbital injection. Thus, emotional reactions were apparently not occurring during the period of complete hemispheric inactivation, but rather during its recovery phase. We have observed some emotional reactions

Table 5-6 Major arterial branches and regions served by anterior, middle, and posterior cerebral arteries

ANTERIOR CEREBRAL ARTERY

Anterior hypothalamus (preoptic and suprachiasmatic nuclei)
Optic chiasm
Region of the optic tract
Septum pellucidum

ACA Branches:
Perforating arteries entering anterior perforated substance
Medial striate artery (anteromedial head of caudate, nearby
    internal capsule and putamen, parts of septal nuclei)
Orbital branches (orbital and medial frontal lobe, part of
    cingulate gyrus)
Frontopolar artery
Callosomarginal artery
Pericallosal artery (ACA's terminal part)

MIDDLE CEREBRAL ARTERY

Lateral portions of orbital gyri
Inferior and middle frontal gyri (lateral convexity)
Large parts of pre- and post-central gyri
Superior and middle temporal gyri including temporal pole
Temporo-occipital, supramarginal, and angular areas

MCA Branches:
Lenticulostriate arteries (anterior perforated substance, head
    of caudate, nearby putamen and internal capsule)
Anterior temporal artery (often anastomoses with
    temporal branches of PCA)
Orbitofrontal artery (may anastomose with ACA)
Pre- and Post-Rolandic branches
Anterior and posterior parietal branches
Posterior temporal branch (lateral occipital region)

Table 5-6 (Continued) Major arterial branches and regions served by anterior, middle, and posterior cerebral arteries

POSTERIOR CEREBRAL ARTERY

Lateral midbrain
Medial and inferior surface of temporal lobe (except temporal pole)
Inferior temporal gyrus
Medial and inferior surface of occipital lobe
Variable portions of lateral occipital lobe
Variable portions of superior parietal lobule
Posterior two-thirds of hippocampus

PCA Branches:
Posterior temporal arteries (occipitotemporal and lingual gyri; inferior lateral surface of temporal lobe; often anastomoses with arteries from MCA)
Posteromedial arteries (mammillary bodies, pituitary gland, infundibulum and tuberal regions of hypothalamus; anterior and medial thalamus; subthalamic region; midbrain; rapheal regions of tegmentum, red nucleus, and medial crus cerebri)
Posterolateral arteries (caudal half of thalamus including geniculate bodies, pulvinar and most lateral nuclear masses)

(Data from Carpenter & Sutin, 1983)

immediately after injection and others developed during the recovery phase suggesting that not all emotional responses following brain damage are due to ipsilateral influences (Lee, Meador, Flanigin & Brooks, 1988; Lee, Loring, Smith & Flanigin, 1990).

Other studies have supported a contralateral release hypothesis where positive (e.g., euphoria, laughter) emotional tendencies intrinsic to the right hemisphere are seen after left hemispheric damage and negative (e.g., depression, crying) emotions intrinsic to the left hemisphere are observed after right hemisphere injury. Sackeim, Greenberg, Weiman, Gur, Hungerbuhler & Geschwind (1982) gathered all available cases of pathological laughing and crying in the neurological literature and found an association between right hemisphere damage and laughing and between left hemisphere damage and crying. Examining only their cases with unilateral lesions, they found pathological laughter in 25 of 33 (76%) cases after right hemisphere damage (8/33 or 24% following left injury) and pathological crying in 16 of 23 (70%) cases after left hemisphere damage (7/23 or 30% showed crying following pright injury). Although these results are congruent with lesion, ECT, and intracarotid amobarbital studies, they do not clarify which side of the brain is disinhibited. That is, these results could be explained by either ipsilateral or contralateral structures driving the emotional reactions.

To determine if the ipsilateral cortex contributes to emotional reactions after brain damage, Sackeim, Greenberg, Weiman, Gur, Hungerbuhler & Geschwind, (1982) went on to examine cases of hemispherectomy. Of 14 right hemispherectomy patients, 12 (86%) were judged euphoric in mood, 1 was judged normal (7%), and 1 was seen as depressed (7%). Because the right cortex was almost completely severed from the rest of the brain, the right cortical centers could not have been responsible for the observed euphoric reactions. There were too few cases of left hemispherectomy to make any firm conclusions about a left hemisphere-depression association. Nevertheless, these results are supportive of the contralateral release hypothesis where the left hemisphere subserves positive emotions and the right hemisphere serves more negative emotions. The hemispherectomy results cannot, however, tell us whether ipsilateral subcortical structures are released from cortical inhibition, thus influencing the type of emotion expressed.

A basis tenet in neurological thinking is that higher brain centers inhibit lower regions. All of the emotional findings above may be explained as the release of subcortical limbic structures from suppression normally exerted downward by the cortex. However, this interpretation does not easily explain the Wada results because the major ipsilateral subcortical structures implicated in emotional expression (e.g., amygdala, anterior hypothalamus, cingulate, orbitofrontal gyrus) which are supplied by the anterior and middle cerebral arteries, are vulnerable to inactivation from the drug. Because amobarbital temporarily disrupts the function of major subcortical limbic structures on the side of injection, it would be difficult to attribute responsibility for specific emotional responses to those barbiturized subcortical limbic structures. Table 5-6 lists the major branching arteries and regions supplied by the anterior, middle, and posterior cerebral arteries. These may be used as a guide to determine which brain structures are affected or unaffected by intracarotid amobarbital administration, although individual variations in cerebral vasculature are common.

The Wada test may be used to shed light on whether there is hemispheric specialization for positive and negative emotions by carefully gauging the time-course of emotional events against the period of drug effect. That is, what cerebral structures are being affected and when?

Transcallosal Inhibition Hypothesis

Converging lines of evidence from lesion, electroconvulsive therapy, and amobarbital studies suggest euphoria most often occurs after right, and depression after left, hemisphere damage. Some have hypothesized that the

Table 5-7 Timing (in Seconds) of emotional reaction onset following amobarbital injection

REACTION	SIDE	DOSE	ONSET
Euphoria/Laughter	Right	100 mg	0
Euphoria/Laughter	Left	100 mg	0
Euphoria/Laughter	Right	125 mg	3
Euphoria/Laughter	Right	100 mg	0
Euphoria/Laughter	Right	100 mg	226
Euphoria/Laughter	Left	150 mg	290
Euphoria/Laughter	Right	150 mg	0
Euphoria/Laughter	Left	250 mg	0
Euphoria/Laughter	Left	125 mg	0
Euphoria/Laughter	Right	100 mg	0
Euphoria/Laughter	Right	100 mg	0
Euphoria/Laughter	Right	175 mg	0
Euphoria/Laughter	Left	100 mg	0
Euphoria/Laughter	Left	200 mg	0
Euphoria/Laughter	Right	100 mg	0
Euphoria/Laughter	Right	100 mg	0
Euphoria/Laughter	Left	125 mg	0
Euphoria/Laughter	Right	200 mg	54
Euphoria/Laughter	Right	100 mg	0
Depression/Crying	Left	150 mg	21
Depression/Crying	Left	100 mg	3
Depression/Crying	Left	200 mg	0
Depression/Crying	Left	125 mg	0

emotion expressed after damage to one hemisphere reflects the intrinsic emotional tone of the opposite, intact hemisphere by release of normal tonic inhibition exerted by 1 hemisphere over the other via the corpus callosum (Flor-Henry, 1979). This "transcallosal inhibition" hypothesis would posit that the left hemisphere is intimately associated with positive emotions such as euphoria and the right hemisphere's intrinsic emotional tone is more negative, being associated with depression for example. In contrast, Tucker (1981) has speculated that emotional response observed after brain damage is caused either by altered functioning of the lesioned hemisphere or by ipsilateral subcortical influences. Thus in Tucker's (1981) system, the opposite hemisphere-emotion association would be posited; viz., the right hemisphere would be specialized for positive (euphoria), and the left for negative (depression), emotions.

If the transcallosal inhibition hypothesis is true, the lesioned hemisphere should be completely inactive and incapable of producing the emotional behavior. To test this hypothesis, we reviewed the videotapes of 23 emotional reactions after amobarbital injection and timed the interval from drug effect to the onset of emotional reactions. If emotional reactions occur shortly after amobarbital administration and before the period of hemispheric anesthesia has elapsed, the brain regions inactivated could not contribute to the genesis of emotional behaviors. Alternatively, if emotional reactions occur during the period of recovery from sedation as Rossi & Rosadini (1967) have anecdotally reported, the injected hemisphere would be capable of influencing the type of emotion expressed.

Twenty (8 male, 12 female) epilepsy surgery candidates who showed a euphoric or depressive emotional reaction after amobarbital injection and who had complete videotaped amobarbital evaluations were selected for independent review. There were 23 emotional reactions, 19 were euphoria/laughter and 4 were depression/crying reactions. The mean time between onset of contralateral hemiparesis (first behavioral evidence of drug effect) and beginning of the emotional reaction was 30 seconds for euphoria and 6 seconds for depression. Of the 19 euphoric responses, 15 began before the onset of hemiparesis. Two of the 4 depressive responses began before hemiparesis onset, while in the 2 remaining cases depression/crying began 3 seconds and 21 seconds after hemiparesis. Results are given in Table 5-7 for each of the 23 emotional reactions.

These results show that emotional reactions following intracarotid amobarbital injection typically begin seconds after drug administration. The interval from first sign of drug effect to the beginning of the emotional response is substantially less than the period of hemispheric inactivation (EEG effects may last 5-7 minutes). Because most emotional reactions begin before the period of hemispheric recovery, and prior to the disappearance of large amplitude, slow EEG waveforms, the cortical and subcortical brain regions affected by amobarbital do not significantly influence the genesis or type of emotional reaction. Our results do not support the previous anecdotal amobarbital emotional research stating that emotional reactions only occur after the motor, sensory, speech, and EEG changes have disappeared, at least 4-6 minutes after injection.

Summary. Our investigation, gauging the onset of emotional responses in relation to which brain structures the amobarbital is affecting, supports the "transcallosal inhibition" hypothesis. Further, the study suggests that there is a positive intrinsic left hemisphere emotional tone and a negative right hemisphere emotional tone. When the territory of the left middle cerebral artery is inactivated, the unopposed "negative/sad" right hemisphere is given free expression. Similarly, the theory supported by this study speculates, albeit less convincingly, that when the right hemisphere is suppressed, the "positive/euphoric" emotional tone of the left hemisphere is released from right hemispheric inhibition via corpus callosum fibers.

Lateralization of Autonomic Function

Emotions have been intimately associated with autonomic functions for over a century (James, 1884), and most definitions of emotion include some reference to autonomic, visceral, or hormonal concommitants to the subjective feeling state. Although there is a rich animal literature regarding higher brain center influences on autonomic functions, the human literature is limited.

Cortical influences on autonomic functions may be mediated by 3 major efferent pathways from orbitofrontal cortex: (1) to the thalamus via the inferior thalamic peduncle, (2) to the temporal lobe and amygdaloid nuclei via the uncinate fasciculus, and (3) to the brainstem through the subthalamus and dorsal hypothalamus (Levine, Patel, Welch & Skinner, 1987). The hypothalamus has terminals in the brainstem nucleus tractus solitarius, cardiovascular motor nuclei, and the nucleus ambiguous. Other cortical connections, such as stria terminalis and diagonal band fibers from the amygdala to hypothalamic and brainstem regions, may also play a role in autonomic control. Cortical areas have been found that influence cardiac function including the anterior tip of the frontal lobe, orbital frontal cortex, motor and premotor cortex, anterior mesial temporal lobe, insula, and cingulate gyrus. Vasomotor responses are abolished by destruction of these cortical regions (Levine, Patel, Welch & Skinner, 1987).

Although asymmetries in human brain function and anatomy are well documented (Geschwind & Galaburda, 1984), data on lateralized differences in cortical control of autonomic function are limited. Heilman, Schwartz & Watson (1978) examined electrodermal responses in patients with unilateral brain damage and normals and found a decreased galvanic skin response (GSR) in right hemisphere damaged patients with unilateral neglect relative to normals and patients with left hemisphere disease. Further, left hemisphere lesioned patients who were aphasic had increased GSRs relative to normals. Yokoyama, Jennings, Ackles, Hood & Boller (1987) studied the effects of unilateral hemisphere lesions on heart rate during a warned reaction time task in which healthy subjects exhibited a normal deceleration in heart rate during the prestimulus period. Consistent with Heilman's findings, they found that right hemisphere lesioned patients had an impairment of anticipatory heart rate deceleration during the prestimulus period of the warned reaction time task. In contrast, left hemisphere lesioned patients showed an accentuated anticipatory deceleration relative to controls.

Zoccolotti, Caltagirone, Benedetti & Gainotti (1986) measured heart rate and galvanic skin responses to video films with negative, neutral, or positive emotional valance in patients and controls with right or left hemisphere lesions. In accordance with other right hemisphere/autonomic function investigations, the authors found that patients with right hemisphere damage exhibited less galvanic skin reactivity and less heart rate deceleration than controls or left hemisphere damaged patients when viewing films with negative emotional content. Rosen, Gur, Sussman, Gur & Hurtig (1982) examined

Table 5-8 Heart rate changes following amobarbital injection

	Left Hemisphere	Right Hemisphere
Baseline	83.6 bpm	83.8 bpm
1st minute	87.4 bpm	83.8 bpm
2nd minute	86.2 bpm	82.3 bpm
3rd minute	85.8 bpm	81.4 bpm

(From Zamrini, Meador, Loring, Nichols, Lee, Figueroa
& Thompson, 1990; adapted with permission of *Neurology*)

heart rate changes in response to intracarotid amobarbital injection in 5 patients. They found that the heart rate increase was greater following left, as compared to right, hemisphere injection in all 5 patients during the first 8 minutes post-injection.

In an expanded study of the effects of unilateral cerebral inactivation on autonomic functioning, we studied heart rate changes following intracarotid amobarbital injection in 25 epilepsy surgery candidates (Zamrini, Meador, Loring, Nichols, Lee, Figueroa & Thompson, 1990). Heart rate increased following left amobarbital injections, but decreased following injection into the right cerebral hemisphere relative to a baseline period (See Table 5-8). Significant differences were present for side of injection and the interaction of side as a function of time. No differences were detected for the order of injection, side of seizure focus, or for differences in drug diffusion to the opposite hemisphere (crossflow). Because the amobarbital doses were so similar in the left (mean = 126 mg) and right (mean = 128 mg) hemispheres, these heart rate differences cannot be due to differences in drug doses delivered to each cerebral hemisphere.

Our heart rate results are congruent with other investigations discussed earlier of autonomic changes after unilateral hemispheric lesions. It seems that autonomic responses may be blunted or inhibited by right hemisphere lesions and exaggerated or disinhibited by left hemisphere lesions. There may be an important mechanism for suppression of autonomic responsiveness and emotional discharge in the anterior regions of the left hemisphere. Left hemisphere inactivation resulted in an abrupt increase in heart rate that tended to drop over 3 minutes. The increase may be due to release of tonic inhibition or to an imbalance in descending influences of the left and right brain on autonomic outflow.

We have previously noted (Lee, Loring, Meador, Flanigin & Brooks, 1988) that severe emotional outbursts that interfere with interpretation of the amobarbital procedure are most likely to occur after right injection among patients with left frontal lesions. Severe outbursts were not seen in patients without left frontal lesions. Other studies have pointed to an important inhibitory role of the left anterior regions of the brain over autonomic/emotional expression. Gainotti (1972) observed that emotional outbursts associated with catastrophic reactions were most frequent in patients with Broca's aphasia. Buck & Duffy (1980) noted that aphasic patients, when viewing emotional slides, were more emotionally expressive than controls, and the degree of emotional expression was correlated with aphasia severity. These authors suggested that the left hemisphere normally exerts an inhibitory influence over emotional expressiveness, and that lesions can result in disinhibition. Heilman, Schwartz & Watson (1978) have also concluded that left hemisphere lesions result in disinhibition of autonomic activity.

In our heart rate study, right hemisphere inactivation resulted in a gradual decline in heart rate consistent with a blunted or diminished responsiveness. Since the right hemisphere appears to have a dominant role in certain aspects of emotional processing and somatic apperception, the reduced heart rate response after right injection may be associated with the right brain damage syndrome consisting of anosognosia, anosodiaphoria, impaired somatosensory processing, and reduced emotional output. Further, the right hemisphere is posited to play a special role in mediating certain aspects of attention and arousal (Huh, Meador, Loring, Lee & Brooks, 1989; Heilman, Watson & Bowers, 1983). This would be consistent with a relatively greater suppression of cortico-reticular networks after right than left brain damage.

Summary and Conclusions

A listing of the various lateralized emotional reactions and autonomic changes that have been reported following unilateral amobarbital injection and after cerebral lesions is given in Table 5-9. Unfortunately, the asymmetries reported in Table 5-9 are not consistent. For instance, asymmetries found in studies of autonomic function are congruent with certain emotional changes after brain damage, but not with others. Specifically, autonomic hyperarousal after left brain damage may be on a continuum with catastrophic reactions, which also have been associated with left lesions. Similarly, autonomic hypoarousal after right brain damage is consistent with emotional changes seen in the indifference reaction, which has in turn, been associated with right lesions. However, hyperarousal is not congruent with depression, sadness, crying, and nonresponsiveness, which also have been correlated with left brain damage. In addition, hypoarousal is not easily conceived of as being similar to euphoria, laughter, and agitation/confusion which have been associated with right brain damage.

Recall that identical autonomic changes may occur in many different emotions and may even be seen in the absence of emotion. Because autonomic changes appear to be a general biological response not specific to any particular emotion, future emotional research will probably not gain much ground by looking for specific types of autonomic differences.

Table 5-9 Lateralized emotional reactions and autonomic changes

	Left Brain	Right Brain
Hécaen (1962): Structural Lesions	Catastrophic Reaction	Indifference
Gainotti (1969): Structural Lesions	Depressive Reaction	Cheerfulness
Robinson et al. (1984): Structural Lesions	Depression	None reported
Sackeim et al. (1982): Structural Lesions	Pathologic Crying	Pathologic Laughter
Sackeim et al. (1982): Hemispherectomy	N too small	Euphoric Mood
Rossi & Rosadini (1967): Amobarbital	Depression	Euphoria
Lee et al. (1990): Amobarbital	Crying	Laughter
Lee et al. (1990): Amobarbital	Nonresponsiveness	Agitation/confusion
Present Chapter: Amobarbital	Positive Emotions	Negative Emotions
Sackeim et al. (1982): Seizures	Gelastic (Laughing)	No Correlation
Heilman et al. (1978): Structural Lesions with GSR	Hyperaroused (GSR)	Hypoaroused
Zamrini et al. (1990): Amobarbital with Heart Rate	Hyperaroused	Hypoaroused
Tucker et al. (1977)	--	Less facial expressiveness
Buck & Duffy (1980)	--	Inaccurate perception of emotion

Left hemisphere damage clearly appears to be associated with depression, sad crying, catastrophic reactions, and hyperarousal of autonomic functions. Right hemisphere damage has consistently been correlated with euphoria, laughing, indifference reactions, and hypoarousal of autonomic functions.

Further, our investigation timing the onset of emotional reactions after intracarotid amobarbital shows the cortical and subcortical regions affected by amobarbital do not substantially influence the genesis or type of emotional reaction. This suggests there is a "positive" intrinsic emotional tone mediated by the left hemisphere and a "negative" right hemisphere emotional tone. When the territory of the internal carotid artery is inactivated, the emotional tone of the unopposed hemisphere is given free expression. Which neural structures and systems within each hemisphere mediate the various emotions remains a mystery.

6
MCG Wada Protocol

Introduction

As with every other epilepsy surgery center, evolutionary changes in our Wada test have occurred. This chapter describes our current protocol employed at the Medical College of Georgia. All patients who are candidates for any type of epilepsy surgery (e.g., temporal lobectomy, corpus callosotomy, focal cortical resection) undergo this procedure.

Patients are informed about the purposes of the testing as well as given specific instructions approximately 1-2 hours prior to the angiographic testing. Patients are explicitly told that some language difficulty is expected, and the patient is told to remember the sentence, "It's a fine day today," and is shown line drawn pictures of a coffee cup and a shoe.

Immediately following angiography, Wada testing is performed with the patient supine. Left and right Wada evaluations are performed on the same day with a minimum of 30 minutes between the 2 injections. Patients are retested prior to the second injection to ensure return to baseline. Sessions are videotaped for subsequent review. The order of amobarbital injection is sequentially alternated. At the onset of testing, patients hold both hands straight up with palms turned rostrally and fingers spread, and begin counting repeatedly from 1-20. This relatively overlearned sequence is employed because in the cases of speech arrest, particularly following right hemisphere injection, we do not infer language if we are able to prompt the patient to repeat counting. Overlearned sequences are easier to resume.

We currently administer a single bolus injection of 100 mg amobarbital sodium (5% solution) by hand via catheter over a 4 second interval following a transfemoral approach into the internal carotid artery. We previously employed an incremental injection approach, increasing dosage initially to a maximum of 250 mg, and later to a maximum of 200 mg, in order to obtain a transient contralateral hemiplegia. However, review of our results revealed a significant relationship between dosage and memory performance (Loring,

Meador & Lee, 1991), with poorer memory associated with higher dosing. Consequently, in order to minimize the number of patients requiring repeat Wada assessment due to poor performance, we now employ low dosage single injections.

Immediately following demonstration of hemiplegia and evaluation of eye-gaze deviation, the patient is requested to execute a simple command (e.g., "touch your nose"). Eight common objects (i.e., "early" items) are then presented for approximately 4 seconds each, and the object names repeated twice to the patient. Examples include a toy shark, pizza cutter, and clothes pin. The patients' eyes are held open as necessary, and the objects are presented in the visual field ipsilateral to the hemisphere injected.

Multiple language tasks are administered. The patient is presented with a modified Token Test in which colored shapes are presented on a vertical card. If the patient cannot execute even a single stage command (e.g., "point to the red circle), we pause in our assessment until some return in language function is observed. We choose to wait until some return of language function prior to continuing our language assessment in order to ensure some similarity to the protocol employed at the Montreal Neurological Institute (Jones-Gotman, 1987). Return of some language function can be demonstrated by the patient's execution of a simple midline command (e.g., "stick out your tongue), and response to simple questions with recognizable, although not necessarily correct, utterances. Most of the language assessment also doubles as memory item presentation. Therefore, we have both early memory items presented during maximum amobarbital effects, and late memory items presented when registration can be assessed.

Two real objects are presented to the patient as a confrontation naming task after some language return has been demonstrated. Paraphasic errors are noted, and if the patient cannot correctly name the object, the correct name is provided by the examiner. Repetition of a simple nursery rhyme is obtained, followed immediately by reading of a simple sentence. Two visual discrimination items from the Florida Kindergarten Screening Battery Recognition-Discrimination subtest are then presented (Satz & Fletcher, 1987). The non-verbal discrimination items are presented to provide balance against the primarily verbal aspects of the other late memory stimuli. The language and discrimination items (2 real objects, nursery rhyme, sentence, 2 visual spatial designs) also serve as memory items for subsequent recall. Patients are informed prior to testing that they would be asked to remember most items presented during the procedure.

Free recall for the 2 line drawings presented prior to the injection, as well as recall for the simple sentence, is obtained. The same pre-injection stimuli are used for both left and right injections, and performance on the test does not affect our inferences regarding memory functioning. If the patient is unable to correctly recall either pre-injection picture, a sequential recognition task is presented. The patient first names each picture, which serves as an additional test of confrontation naming ability, and then indicates if that picture had been previously seen. Four foils are included. If the patient is able to freely recall the 2 pre-injection pictures, confrontation naming is still obtained, but the patients are not asked if they had previously seen the stimuli. No recognition for the pre-injection sentence is obtained if the patient cannot freely recall it. The recognition memory foils serve as additional memory stimuli to be recognized following return to baseline functioning.

Language Rating

Formal language rating is based upon performance on 4 linguistic tasks (viz., counting disruption, comprehension, naming, and repetition). Although we have not yet formally incorporated other language tasks into our rating scale, the presence of any aphasic error during other portions of the test are considered positive signs of language representation, and our clinical decisions are weighed accordingly.

The expressive language score (0-4) is based upon disruption of counting ability (0 = normal, slowed, or brief pause ≤ 20 seconds; 1 = counting perseveration with normal sequencing; 2 = sequencing errors; 3 = single number or word perseveration; 4 = arrest > 20 seconds). Comprehension from the modified Token Test is rated on a 4-point scale: 1. "point to the red circle after the green square," 2. "point to the red circle and then point to the green square," 3. "point to the red triangle." A score of 0 is awarded for completion of the complex 2-stage command with inverted syntax, a score of 1 reflects successful simple 2-stage command, 2 is scored for the 1-stage commands, and 3 if the subject cannot perform any commands. Confrontation naming for the 2 objects is scored as pass or fail for each stimulus. Nursery rhyme repetition is graded on a 0-3 rating scale. In all 4 categories, a score of 0 reflects normal function.

Language classification is determined by performance on each of the 4 language categories assessed, viz., counting disruption, aural comprehension, naming, and repetition. Because considerable differences exist regarding reaction to the medication and duration of the anaesthetic effects, we adopted a conservative classification for language representation. For language impairment to be inferred, as distinct from confusion or abulia, 1 of 2 error configurations had to be detected. In the first, impairments (scores > 0) had to be present in at least 2 categories, with 1 of the scores greater than 1. In the second pattern, language representation would be inferred if at least 3/4 language categories are only mildly impaired (e.g., scores of 1). We should emphasize that these are our current research criteria for inferring the presence of language representation, and if any clinical doubt exists regarding the presence of language representation in a hemisphere in which surgery will be performed, and review of the videotape is insufficient to remove this doubt, cortical speech mapping will be performed during the operation.

Memory Testing

Recognition memory of material presented during the procedure is tested after the amobarbital effects have worn off, as demonstrated by complete return of 5/5 strength, absence of pronator drift, absence of bradykinesia and asterixis, normal repetition of the phrases "no ifs, ands, or buts," and "Methodist Episcopal," as well as the ability to execute complex 2-stage commands involving inverted syntax, To assess early object memory, the 8 early target items are presented individually with 16 randomly interspersed foils. The subject indicates if each item has been previously presented. To correct for response bias and guessing, we incorporated a correction by subtracting from the total correct, 1/2 the incorrect (false positive) responses. In practice, however, the number of false positive responses tends to be small in TLE patients, and in very few patients does this adjustment appreciably change a patient's score. Using our formula, any average of guessing without recognition would yield an expected value of 0 [e.g., (8 correct) - 0.5 x (16 incorrect) = 0]. This is our primary memory measure used to infer or to discount risk of post-surgical amnesia. We do not use firm pass/fail criteria, but generally, if the patient does not recognize at least 2 early objects following injection ipsilateral to the side of contemplated surgery, we will repeat the test on this side.

Recognition tasks are also used to assess memory for the stimuli presented after at least some return of language functions ("late" items). If unable to freely recall the 2 objects used for confrontation naming, which is typical given the interference from the early item stimuli, the names of 5 objects are read to the patient, and the patient is instructed to identify those objects presented. The number of targets is not indicated. Similarly, if unable to freely recall the rhyme, 4 nursery rhymes are read to the patient for recognition assessment. Only recognition assessment is analyzed, and free recall is scored as evidence of correct recognition. A multiple choice recognition assessment is obtained for the 2 recognition-discrimination geometric designs, with the targets interspersed on the same recognition sheet with 4 foils. Again, the number of targets is not indicated, and the subject is asked following each response, either correct or incorrect, if any additional items are recognized. Although the same ratio of targets to foils does not apply given the different recognition formats employed, we still correct for guessing by subtracting 1/2 the number of false positive responses from the sum of the correctly recognized items.

Multiple choice recognition for the sentence is obtained, as well as a sequential recognition for the foils for the pre-injection recognition tasks. As with the other late item memory performances, these are considered secondary measures for individual patient evaluation. They primarily serve to increase or decrease the likelihood of repeat Wada testing. For example, correct recognition of 2 early items ipsilateral to the seizure focus might be sufficient with good late item recall, qualitatively determined. However, this same score might be insufficient to proceed to surgery without repeated Wada

evaluation if poor late item performance is present. An outline of our scoring sheet is presented on pages 114-117.

WADA EVALUATION

Name: _____   MCG: _____   Date: _____   Age: _____
Sex: _____   Handedness: _____   Rater: _____

INJECTION 1 Right Left Dose (mg) _____   Flaccid Hemiparesis Y N
Time Test Began: _____

Initial Expressive Language (Counting)   0 1 2 3 4   0 sec   Running _____
                                                                                            Time _____

0=normal, slowed or brief pause; 1=counting perseveration with normal sequencing;
2=sequencing errors; 3=single number or word perseveration; 4=arrest

0=normal, 1=mild
2=moderate, 3=severe

Lateral Gaze Palsy	None	R	L
Spontaneous	___	___	___	0 1 2 3
Command	___	___	___	0 1 2 3 N/A

RECEPTIVE LANGUAGE (Initial Simple):   0 1 2 3   Time _____

Modified Glascow Coma Scale   If ptosis, note severity: 0 1 2 3 (0=normal)
                                          If akinesia, note severity: 0 1 2 3

Eye Opening: Eye Movements:

Motor (ipsilateral arm):   C
      (contralateral arm):

	C		
spontaneous	6	6	spontaneous	4	follows objects well	4
localizes	5	5	to loud noise	3	follows objects variably	3
withdraws	4	4	to pain	2	orients to loud noise	2
abnormal flexion	3	3	nil	1	and light	2
extensor response	2	2			nil	1
nil	1					

EARLY ITEM PRESENTATION:

Time Begin _____
Time End _____

COMPREHENSION (Delayed-Token card):   0 1 2 3
(0=normal, 1=mild, 2=moderate, 3=severe)

EXPRESSIVE LANGUAGE:

Naming (screwdriver, pipe)   (errors) 0 1 2
    Errors: _____
Repetition (Mary had a little lamb)   0 1 2 3 N/A
Dysarthria   0 1 2 3 N/A
Reading (The car backed over the curb.)   0 1 2 3 N/A

VISUAL DISCRIMINATION   (errors) 0 1 2

PREINJECTION ITEM RECALL:

Picture Free Recall   (errors) 0 1 2
Picture Recognition   (errors) 0 1 2   N/A

Set A: pen _____ Y N   pear _____ Y N   FP _____
       broom _____ Y N   shoe _____ Y N   Time _____
       cup _____ Y N   lamp _____ Y N   FP _____
                             FP _____

Sentence Free Recall   Y N FP _____   Time _____

RETEST:

Strength _____   Time _____
Repetition (No ifs ands or buts; Methodist-Episcopal)   0 1 2 3
Comprehension   0 1 2 3

Time at which strength and language normal _____   Time _____

Time _____

EARLY OBJECT RECOGNITION:

___ Y N	___ Y N	___ Y N
___ Y N	___ Y N	___ Y N
___ Y N	___ Y N	___ Y N
___ Y N	___ Y N	___ Y N
___ Y N	___ Y N	___ Y N
___ Y N	___ Y N	___ Y N

Targets

List 1:   shark, helicopter, ball, car, boat, tongs, quarter, eraser

List 2:   clothes pin, scrub pad, bowl, pizza cutter, battery, shell, toothbrush, rope

LATE ITEM RECALL/RECOGNITION: Time _____

Rhyme Free Recall   Y N   FP ___

Rhyme Recognition   Y N   N/A   FP ___

A=Old Mother Hubbard went to the cupboard
B=Mary had a little lamb
C=Mary, Mary Quite Contrary
D=Three blind mice, Three blind mice

Object Free Recall   (errors) 0 1 2   FP ___

Object Recognition   (errors) 0 1 2   N/A

(key, pipe, hammer, paper clip, screwdriver)   FP ___

Visuospatial Recognition   (errors) 0 1 2   FP ___

Sentence Recognition   Y N   N/A   FP _____

1. The car came to a sudden stop.
2. The boat sailed across the lake.
3. The car backed over the curb.
4. The airplane rose above the clouds.

Card Recognition

    Set A: telephone _____ Y N bird _____ Y N
                sword _____ Y N piano _____ Y N
                apple _____ Y N bed _____ Y N
                lamp _____ Y N pear _____ Y N
                cow _____ Y N bridge _____ Y N
                pen _____ Y N broom _____ Y N

Correct: _____

FP _____

Denial/Anosognosia: 0 1

Comments: _____

Time Completed: _____

Affect Change:
Response Perseveration:
Paraphasic Errors:
Attentional Deficits:
Ptosis:

References

Aasly, J., & Silfvenius, H. (1990). Evaluation of early and late presented tasks in the intracarotid Amytal test for epileptic patients. *Epilepsy Research, 7*, 155-164.

Ajersch, M. K., & Milner, B. (1983). Handwriting posture as related to cerebral speech lateralization, sex, and writing hand. *Human Neurobiology, 2*, 143-145.

Alema, G., & Donini, G. (1960). Sulle modificazioni cliniche ed elettroencefalografiche da introduzione intracarotidea di iso-amil-etil-barbiturato di sodio nell'uomo. *Bollettino-Societa Italiana Biologia Sperimentale, 36*, 900-904.

Alema, G., & Rosadini, G. (1964). Donnees cliniques et E.E.G. de l'introduction d'Amytal sodium dans la circulation encephalique, concernant l'etat de conscience. *Acta Neurochirurgica, 12*, 240-257.

Andral, G. (1834). Observations sur les maladies de l'encéphale et de ses enveloppes. In *Clinique medicalé au choix d'observations rescueilliés à la clinique de M. Lerminier*, Vol. 5, 3rd ed. Paris: D. Cazellin.

Annett, M. (1975). Hand preference and the laterality of cerebral speech. *Cortex, 11*, 305-328.

Baldwin, M. (1956). Modifications psychiques survenant apré lobectomie temporale subtotale. *Neurochirurgie, 2*, 152-167.

Bender, M. B. (1980). Brain control of conjugate horizontal and vertical eye movements: A survey of the structural and functional correlates. *Brain, 103*, 23-69.

Benson, D. F. (1985). Left-hemisphere language. In D. F. Benson & E. Zaidel (Eds.), *The dual brain: Hemispheric specialization in humans* (pp. 193-203). New York: The Guilford Press.

Benson, D. F., & Geschwind, N. (1985). Aphasia and related disorders: A clinical approach. In M-M. Mesulam (Ed.), *Principles of behavioral neurology* (pp. 193-238). Philadelphia: FA Davis.

Bizzi, E., Kalil, R.E., & Tagliasco, V. (1971). Eye-head coordination in monkeys: Evidence for centrally patterned organization. *Science, 173,* 452-454.

Blume, W. T., Grabow, J. D., Darley, F. L., & Aronson, A. E. (1973). Intracarotid amobarbital test of language and memory before temporal lobectomy for seizure control. *Neurology, 23,* 812-819.

Branch, C., Milner, B., & Rasmussen, T. (1964). Intracarotid Sodium Amytal for the lateralization of cerebral speech dominance: Observations in 123 patients. *Journal of Neurosurgery, 21,* 399-405.

Bryden, M. P., & Rainey, C. A. (1963). Left-right differences in tachistoscopic recognition. *Journal of Experimental Psychology, 66,* 568-571.

Buck, R., & Duffy, R. J. (1980). Nonverbal communication of affect in brain damaged patients. *Cortex, 16,* 351-362.

Campbell, R. (1978). Asymmetries in interpreting and expressing a posed facial expression. *Cortex, 14,* 327-342.

Cannon, W. B. (1927). The James-Lange theory of emotions. *American Journal of Psychology, 39,* 106-124.

Carmon, A., & Benton, A. L. (1969). Tactile perception of direction and number in patients with unilateral cerebral disease. *Neurology, 19,* 525-532.

Carpenter, M. B., & Sutin, J. (1983). *Human neuroanatomy,* 8th ed. Baltimore: Williams and Wilkins.

Castro-Caldas, A., Confraria, A., & Poppe, P. (1987). Non-verbal disturbances in crossed aphasia. *Aphasiology, 1,* 403-413.

Christianson, S. A., Säisä, J., & Sifvenius, H. (1990). Hemisphere memory differences in Sodium Amytal testing of epileptic patients. *Journal of Clinical and Experimental Neuropsychology, 12,* 681-694.

Christianson, S. A., Silfvenius, H., & Nilsson, L. G. (1987). Hemisphere memory of concrete and abstract information determined with the intracarotid Sodium Amytal test. *Epilepsy Research*, *1*, 185-193.

Corkin, S. (1984). Lasting consequences of bilateral medial temporal lobectomy: Clinical course and experimental findings in H. M. *Seminars in Neurology*, *4*, 249-258.

Cummings, J. L., Tomiyasu, U., Read, S., & Benson, D. F. (1984). Amnesia with hippocampal lesions after cardiopulmonary arrest. *Neurology*, *34*, 679-681.

Damasio, A., Bellugi, U., Damasio, H., Poizner, H., & Van Gilder, J. (1986). Sign language aphasia during left-hemisphere Amytal injection. *Nature*, *322*, 363-365.

Deglin, V. L., & Nikolaenko, N. N. (1975). Role of the dominant hemisphere in the regulation of emotional states. *Human Physiology*, *1*, 394-402.

De Renzi, E., Faglioni, P., & Scotti, G. (1970). Hemispheric contribution to the exploration of space through the visual and tactile modality. *Cortex*, *6*, 191-203.

Desmedt, J. E. (1977). Active touch exploration of extrapersonal space elicits specific electrogenesis in the right cerebral hemisphere of intact right handed man. *Procedures of the National Academy of Science in the USA*, *74*, 4037-4040.

DeToledo, J., Smith, D. B., Kramer, R. E. (1989). A proposed cause for the variability of memory lateralization in Amytal suppression (Wada) testing: Role of the septal region. *Epilepsia*, *30*, 712. (Abstract).

Dimsdale, H., Logue, V., & Piercy, M. (1964). A case of persisting impairment of recent memory following right temporal lobectomy. *Neuropsychologia*, *1*, 287-298.

Dinner, D. S., Lüders, H., Morris III, H. H., Wyllie, E., & Kramer, R. E. (1987). Validity of intracarotid sodium amobarbital (Wada test) for evaluation of memory function. *Neurology*, *37*(Suppl. 1), 142. (Abstract)

Engel, J., Jr. (1987). *Surgical treatment of the epilepsies*. New York: Raven Press.

Engel, J., Jr., Crandall, P. H., & Rausch, R. (1983). The partial epilepsies. In R. N. Rosenberg & R. G. Grossman (Eds.), *The clinical neurosciences* (pp. II:1349-II:1380). New York, Edinburgh, London, and Melbourne: Churchill Livingstone.

Engel, J., Jr., Rausch, R., Lieb, J. P., Kuhl, D. E., & Crandall, P. H. (1981). Correlation of criteria used for localizing epileptic foci in patients considered for surgical therapy of epilepsy. *Annals of Neurology, 9,* 215-224.

Faglioni, P., Scotti, G. & Spinnler, H. (1971). The performance of brain-damaged patients in spatial localization of visual and tactile stimuli. *Brain, 94,* 443-454.

Fedio, P., & Weinberg, L. K. (1971). Dysnomia and impairment of verbal memory following intracarotid injection of Sodium Amytal. *Brain Research, 31,* 159-168.

Flanigin, H. F., Schlosberg, A., Power, J., & Smith. J. (1985). Evaluation of memory by localized intraoperative cooling. *Epilepsia, 26,* 543. (Abstract).

Flor-Henry, P. (1979). On certain aspects of the localization of the cerebral systems regulating and determining emotion. *Biological Psychiatry, 4,* 677-698.

Gainotti, G. (1969). Réactions catastrophiques et manifestations d'indifférence au cours des atteintes cerebrales. *Neuropsychologia, 7,* 195-204.

Gainotti, G. (1972). Emotional behavior and hemispheric side of lesion. *Cortex, 8,* 41-55.

Gainotti, G. (1983). Laterality of affect: The emotional behavior of right- and left-brain-damaged patients. In M. S. Myslobodsky (Ed.), *Hemisyndromes: Psychobiology, neurology, psychiatry* (pp. 175-192). New York: Academic Press.

Gardner, W. J. (1941). Injection of procaine into the brain to locate speech area in left-handed persons. *Archives of Neurology and Psychiatry, 46,* 1035-1038.

Gasparrini, W. G., Satz, P., Heilman, K. M., & Coolidge, F. L. (1978). Hemispheric asymmetries of affective processing as determined by the Minnesota Multiphasic Personality Inventory. *Journal of Neurology, Neurosurgery, and Psychiatry, 41,* 470-473.

Geschwind, N., & Galaburda, A. M. (Eds.). (1984). *Cerebral dominance: The biological foundations*. Cambridge, MA: Harvard University Press.

Gilman, S., MacFadyen, D. J., & Denny-Brown, D. (1963). Decerebrate phenomena after carotid amobarbital injection. *Archives of Neurology, 8,* 662-675.

Girvin, J. P., McGlone, J., McLachlan, R. S., & Blume, W. T. (1987). Validity of the sodium amobarbital test for memory in selected patients. *Epilepsia, 28,* 636. (Abstract).

Gloning, K. (1977). Handedness and aphasia. *Neuropsychologia, 15,* 355-358.

Goldberg, M. E., & Segraves, M. A. (1987). Visuospatial and motor attention in the monkey. *Neuropsychologia, 25,* 107-118.

Hart, J., Jr., Lesser, R. P., Fisher, R. S., Schwerdt, P., Bryan, R. N., & Gordon, B. (1991). Dominant-side intracarotid amobarbital spares comprehension of word meaning. *Archives of Neurology, 48,* 55-58.

Hécaen, H. (1962). Clinical symtomatology in right and left hemispheric lesions. In V. B. Mountcastle (Ed.), *Interhemispheric relations and cerebral dominance* (pp. 215-243). Baltimore: Johns Hopkins Press.

Hécaen, H., Mazars, G., Ramier, A. M., Goldblum, M. C., & Mérienne, L. (1971). Aphasie croisée chez un sujet droitier bilingue. *Revue Neurologique, 1,* 319-323.

Heilman, K. M., Bowers, D., Coslett, H., Whelan, H., & Watson, R. (1985). Directional hypokinesia: Prolonged reaction times for leftward movements in patients with right hemisphere lesions and neglect. *Neurology, 35,* 855-859.

Heilman, K. M., Bowers, D., & Valenstein, E. (1985). Emotional disorders associated with neurological diseases. In K. M. Heilman & E. Valenstein (Eds.), *Clinical Neuropsychology*, 2nd ed. (pp. 377-402). New York: Oxford University Press.

Heilman, K. M., & Gonzales Rothi, L. J. (1985). Apraxia. In K. M. Heilman & Valenstein, E. (Eds.), *Clinical neuropsychology*, 2nd ed. (pp. 131-150). New York: Oxford University Press.

Heilman, K. M., Schwartz, H., & Watson, R. T. (1978). Hypoarousal in patients with the neglect syndrome and emotional indifference. *Neurology, 28,* 229-232.

Heilman, K. M., & Valenstein, E. (1979). Mechanisms underlying hemispatial neglect. *Annals of Neurology*, *5*, 166-170.

Heilman, K. M., & Valenstein, E. (Eds.) (1985). *Clinical neuropsychology*, 2nd ed. New York and Oxford: Oxford University University Press.

Heilman, K. M., Valenstein, E., & Watson, R. T. (1985). The neglect syndrome. In P. J. Vinken, G. W. Bruyn, H. L. Klawans, & J. A. M. Frederis (Eds.), *Handbook of clinical neurology: Vol. 45. Clinical neuropsychology* (pp. 153-184). Amsterdam: Elsevier Science.

Heilman, K. M., & Van Den Abell, T., (1980). Right hemispheric dominance for attention: The mechanism underlying hemispheric asymmetries of inattention (neglect). *Neurology*, *30*, 327-330.

Heilman, K. M., Watson, R. T., & Bowers, D. (1983). Affective disorders associated with hemispheric disease. In K. M. Heilman & P. Satz (Eds.), *Neuropsychology of human emotion* (pp. 45-64). New York: Guilford Press.

Hietala, S. O., Silfvenius, H., Aasly, J., Olivercrona, M., & Jonsson, L. (1990). Brain perfusion with intracarotid injection of 99mTc-HM-PAO in partial epilepsy during amobarbital testing. *European Journal of Nuclear Medicine*, *16*, 683-687.

Hommes, O. R., & Panhuysen, L. H. H. M. (1970). Bilateral intracarotid Amytal injection: A study of dysphasia, disturbance of consciousness and paresis. *Psychiatry, Neurology, and Neurochirurgia*, *73*, 447-459.

Horel, J. A. (1978). The neuroanatomy of amnesia: A critique of the hippocampal memory hypothesis. *Brain*, *101*, 403-445.

Huh, K., Meador, K. J., Loring, D. W., Lee, G. P., & Brooks, B. S. (1989). Attentional mechanisms during the intracarotid amobarbital test. *Neurology*, *39*, 1183-1186.

Jack, C. R., Jr., Bentley, M. D., Twomey, C. K., & Zinsmeister, A. R. (1990). MR imaging-based volume measurements of the hippocampal formation and anterior temporal lobe: Validation studies. *Radiology*, *176*, 205-209.

Jack, C. R., Jr., Nichols, D. A., Sharbrough, F. W., Marsh, W. R., & Petersen, R. C. (1988). Selective posterior cerebral artery Amytal test for evaluating memory function before surgery for temporal lobe seizure. *Radiology*, *168*, 787-793.

Jack, C. R., Jr., Sharbrough, F. W., Twomey, C. K., Cascino, G. D., Hirschorn, K. A., Marsh, W. R., Zinsmeister, A. R., & Scheithauer, B. (1990). Temporal lobe seizures: Lateralization with MR volume measurements of the hippocampal formation. *Radiology, 175*, 423-429.

James, W. (1884). What is emotion? *Mind, 19*, 188-205.

Jeffery, P. J., Monsein, L. H., Szabo, Z., Hart, J., Fisher, R. S., Lesser, R. P., Debrun, G. M., Gordon, B., Wagner, Jr., H. N., & Camargo, E. E. (1991). Mapping the distribution of amobarbital sodium in the intracarotid Wada test by use of Tc-99m HMPAO with SPECT. *Radiology, 178*, 847-850.

Jones-Gotman, M. (1987). Commentary: Psychological evaluation--Testing hippocampal function. In J. Engel, Jr. (Ed.), *Surgical treatment of the epilepsies* (pp. 203-211). New York: Raven Press.

Kimura, D. (1961). Cerebral dominance and the perception of verbal stimuli. *Canadian Journal of Psychology, 15*, 166-171.

Kløve, H., Grabow, J. D., & Trites, R. L. (1969). Evaluation of memory functions with intracarotid sodium Amytal. *Transactions of the American Neurological Association, 94*, 76-80.

Kløve, H., Trites, R. L., & Grabow, J. D. (1970). Intracarotid sodium Amytal for evaluating memory function. *Electroencephalography and Clinical Neurophysiology, 28*, 418-419. (Abstract No. 7).

Kurthen, M., Linke, D. B., Reuter, B. M., Hufnagel, A., & Elger, C. E. (1991). Severe negative emotional reactions in intracarotid sodium amytal procedures: Further evidence for hemispheric asymmetries? *Cortex, 27*, 333-337.

Kurthen, M., Reichmann, K., Linke, D. B., Biersack, H. J., Reuter, B. M., Durwen, H. F., & Grünwald, F. (1990). Crossed cerebellar diaschisis in intracarotid sodium amytal procedures: A SPECT study. *Acta Neurologica Scandinavica, 81*, 416-422.

Lansdell, H. (1969). Verbal and nonverbal factors in right-hemisphere speech: Relation to early neurological history. *Journal of Comparative and Physiological Psychology, 69*, 734-738.

Larrabee, G. J., Kane, R. L., & Rogers, J. A. (1982). Neuropsychological analysis of a case of crossed aphasia: Implications for reversed laterality. *Journal of Clinical Neuropsychology, 4*, 131-142.

Lee, G. P., Loring, D. W., Meador, K. J., & Brooks, B. B. (1990). Hemispheric specialization for emotional expression: A reexamination of results from intracarotid administration of sodium amobarbital. *Brain and Cognition, 12,* 267-280.

Lee, G. P., Loring, D. W., Meador, K. J., Flanigin, H. F., & Brooks, B. S. (1988). Severe behavioral complications following intracarotid sodium amobarbital injection: Implications for hemispheric asymmetry of emotion. *Neurology, 38,* 1233-1236.

Lee, G. P., Loring, D. W., Smith, J. R., & Flanigin, H. F. (1990). Material specific learning during electrical stimulation of the human hippocampus. *Cortex, 26,* 433-442.

Lesser, R. P., Dinner, D. S., Lüders, H., & Morris, H. H. (1986). Memory for objects presented soon after intracarotid amobarbital sodium injections in patients with medically intractable complex partial seizures. *Neurology, 36,* 895-899.

Levine, S. R., Patel, V. M., Welch, K. M. A., & Skinner, J. E. (1987). Are heart attacks really brain attacks? In J. J. Furlan (Ed.), *The heart and stroke: Exploring mutual cerebrovascular and cardiovascular issues* (pp. 185-216). London: Springer-Verlag.

Ley, R. G., & Bryden, M. P. (1979). Hemispheric differences in recognizing faces and emotions. *Brain and Language, 7,* 127-138.

Loring, D. W., Flanigin, H. F., Meador, K. J., Lee, G. P., & Smith, J. R. (1988). Right hemisphere language representation determined by intracarotid sodium amytal testing and functional cortical speech mapping. *Epilepsia, 29,* 686. (Abstract).

Loring, D. W., Lee, G. P., Flanigin, H. F., Meador, K. J., Smith, J. R., Gallagher, B. B., & King, D. W. (1988). Verbal memory performance following unilateral electrical stimulation of the human hippocampus. *Journal of Epilepsy, 1,* 79-85.

Loring, D. W., Lee, G. P., & Meador, K. J. (1989). The intracarotid amobarbital sodium procedure: False-positive errors during recognition memory assessment. *Archives of Neurology, 46,* 285-287.

Loring, D. W., Lee, G. P, Meador, K. J., Flanigin, H. F., Smith, J. R., Figueroa, R. E., & Martin, R. C. (1990). The intracarotid amobarbital procedures as a predictor of memory failure following unilateral temporal lobectomy. *Neurology, 40,* 605-610.

Loring, D. W., Meador, K. J., & Lee, G. P. (1989). Effects of seizure focus on cerebral motor organization as determined by intracarotid amobarbital testing. *Epilepsia, 30,* 677. (Abstract).

Loring, D. W., Meador, K. J., & Lee, G. P. (1990). Should memory deficits following unilateral temporal lobectomy be considered amnesia? A critical review of the original literature. *International Cleveland Clinic Symposium Bulletin: Epilepsy Surgery, 2,* 52. (Abstract).

Loring, D. W., Meador, K. J., & Lee, G. P. (1991). Amobarbital dosage effects on Wada memory. *Neurology, 41,* 299. (Abstract).

Loring, D. W., Meador, K. J., Lee, G. P., Flanigin, H. F., King, D. W., & Smith, J. R. (1990). Crossed aphasia in a patient with complex partial seizures: Evidence from intracarotid amobarbital testing, functional cortical mapping, and neuropsychological assessment. *Journal of Clinical and Experimental Neuropsychology, 12,* 340-354.

Loring, D. W., Meador, K. J., Lee, G. P., & Martin, R. C. (1990). Unilateral amobarbital injections fail to produce retrograde memory deficits. *Journal of Clinical and Experimental Neuropsychology, 12,* 86. (Abstract).

Loring, D. W., Meador, K. J., Lee, G. P., Murro, A. M., Smith, J. R., Flanigin, H. F., Gallagher, B. B., & King, D. W. (1990). Cerebral language lateralization: Evidence from intracarotid amobarbital testing. *Neuropsychologia, 28,* 831-838.

Mateer, C. A., & Dodrill, C. B. (1983). Neuropsychological and linguistic correlates of atypical language lateralization: Evidence from sodium amytal studies. *Human Neurobiology, 2,* 135-142.

McGlone, J. (1984). Speech comprehension after unilateral injection of sodium Amytal. *Brain and Language, 22,* 150-157.

McGlone, J., & MacDonald, B. H. (1989). Reliability of the sodium amobarbital test for memory. *Journal of Epilepsy, 2,* 31-39.

Meador, K. J., Loring, D. W., Lee, G. P., Brooks, B. S., Nichols, F. T., Thompson, E. E., Thompson, W. O., & Heilman, K. M. (1989). Hemisphere asymmetry for eye gaze mechanisms. *Brain, 112,* 103-111.

Meador, K. J., Loring, D. W., Lee, G. P., Brooks, B. S., Thompson, E. E., Thompson, W. O., & Heilman, K. M. (1988). Right cerebral specialization for tactile attention as evidenced by intracarotid sodium amytal. *Neurology, 38,* 1763-1766.

Mesulam, M-M. (1981). A cortical network for directed attention and unilateral neglect. *Annals of Neurology, 10*, 309-325.

Milner, B. (1966). Amnesia following operation on the temporal lobes. In C. W. M. Whitty & O. L. Zangwill (Eds.), *Amnesia* (pp. 109-133). London: Butterworths.

Milner, B. (1969). Evaluation of memory functions with intracarotid sodium amytal: Discussion. *Transactions of the American Neurological Association, 94*, 79-80.

Milner, B. (1972). Disorders of learning and memory after temporal lobe lesions in man. *Clinical Neurosurgery, 19*, 421-446.

Milner, B. (1975). Psychological aspects of focal epilepsy and its neurosurgical management. In D. Purpura, J. Penry, & R. Walter (Eds.), *Advances in neurology: Vol. 8. Neurosurgical management of the epilepsies* (pp. 299-321). New York: Raven Press.

Milner, B., & Branch, C. (1967). In G. F. Rossi & G. Rosadini, Experimental analysis of cerebral dominance in man. In C. H. Millikan and F. L. Darley (Eds.), *Brain mechanisms underlying speech and language* (pp. 177-184). New York: Grune & Stratton.

Milner, B., Branch, C., & Rasmussen, T. (1962). Study of short-term memory after intracarotid injection of sodium Amytal. *Transactions of the American Neurological Association, 87*, 224-226.

Milner, B., Branch, C., & Rasmussen, T. (1966). Evidence for bilateral speech representation in some non-right-handers. *Transactions of the American Neurological Association, 91*, 306-308.

Mohr, J. P., Rubinstein, L. V, Kase, C. S., Price, T. R., Wolfe, P. A., Nichols, F. T., & Tatemichi, T. K. (1984). Gaze palsy in hemispheral stroke: The NINCDS Stroke Data Bank. *Neurology, Cleveland, 34*(Suppl. 1), 199.

Moster, M., & Goldberg, G. (1990). Topography of scalp potentials preceding self-initiated saccades. *Neurology, 40*, 644-648.

Nichols, F. T., Mawad, M. E., Hilal, S. K., Mohr, J. P., Michelsen, W. J., & Stein B. M. (1985). Headache patterns and transient neurologic deficits during temporary balloon occlusion of intracranial arteries. *Stroke, 16*, 149.

Novelly, R. A., & Naugle, R. I. (1985). Gender specific effects on VIQ versus PIQ in acquired right-hemisphere speech. *Journal of Clinical and Experimental Neuropsychology*, 7, 627. (Abstract No. 10)

Novelly, R. A., & Williamson, P. D. (1989). Incidence of false-positive memory impairment in the intracarotid Amytal procedure. *Epilepsia*, 30, 711. (Abstract)

Oxbury, S. M., & Oxbury, J. M. (1984). Intracarotid amytal test in the assessment of language dominance. In F. C. Rose (Ed.), *Advances in neurology: Vol. 42. Progress in aphasiology* (pp. 115-123). New York: Raven Press.

Penfield, W., & Mathieson, G. (1974). Memory: Autopsy findings and comments on the role of the hippocampus in experimental recall. *Archives of Neurology*, 31, 145-154.

Penfield, W., & Milner, B. (1958). Memory deficit produced by bilateral lesions in the hippocampal zone. *Archives of Neurology and Psychiatry*, 79, 475-497.

Perria, L., Rosadini, G., & Rossi, G. F. (1961). Determination of side of cerebral dominance with amobarbital. *Archives of Neurology*, 4, 173-181.

Plum, F., & Posner, J. B. (1983). *The diagnosis of stupor and coma*, 3rd ed. Philadelphia: FA Davis.

Powell, G. E., Polkey, C. E., & Canavan, A. G. M. (1987). Lateralisation of memory functions in epileptic patients by use of the sodium amytal (Wada) technique. *Journal of Neurology, Neurosurgery, and Psychiatry*, 50, 665-672.

Press, G. A., Amaral, D. G., & Squire, L. R. (1989). Hippocampal abnormalities in amnestic patients revealed by high-resolution magnetic resonance imaging. *Nature*, 341, 54-57.

Prevost, J-L. (1868). *De la déviation conjuguée des yeux et de la rotation de la tête dans certains cas d'hémiplégie*. Thesis, Paris, No. 30.

Pribram, K. H. & Melges, F. T. (1969). Psychophysiological basis of emotion. In P. J. Vinken & A. W. Bruyn (Eds.), *Handbook of clinical neurology, Vol. 3* (pp. 316-342). Amsterdam: Elsevier/North-Holland.

Pycock, C. J. (1980). Turning behavior in animals. *Neuroscience*, 5, 461-514.

Rasmussen, T., & Milner, B. (1975). Clinical and surgical studies of the cerebral speech areas in man. In K. J. Zülch, O. Creutzfeldt, & G. C. Galbraith (Eds.), *Cerebral localization: An Otfrid Foerster symposium* (pp. 238-257). New York, Heidelberg, Berlin: Springer-Verlag.

Rasmussen, T., & Milner, B. (1977). The role of early left-brain injury in determining lateralization of cerebral speech functions. *Annals of the New York Academy of Science, 299*, 355-369.

Rausch, R. (1987). Psychological evaluation. In J. Engel, Jr. (Ed.), *Surgical treatment of the epilepsies* (pp. 181-195). New York: Raven Press.

Rausch, R., Babb, T. L., Engel, J., & Crandall, P. H. (1989). Memory following intracarotid amobarbital injection contralateral to hippocampal damange. *Archives of Neurology, 46*, 783-788.

Rausch, R., Boone, K., & Ary, C. M. (1991). Right-hemisphere language dominance in temporal lobe epilepsy: Clinical and neuropsychological correlates. *Journal of Clinical and Experimental Neuropsychology, 13*, 217-231.

Rausch, R., Fedio, P., Ary, C. M., Engel, J., Jr., & Crandall, P. H. (1984). Resumption of behavior following intracarotid sodium amobarbital injection. *Annals of Neurology, 15*, 31-35.

Rausch, R., & Walsh, G. (1984). Right-hemisphere language dominance in right-handed epileptic patients. *Archives of Neurology, 41*, 1077-1080.

Reivich, M., Alavi, A., & Gur, R. C. (1984). Positron emission tomographic studies of perceptual tasks. *Annals of Neurology, 15*(Suppl.), S61-S65.

Rey, M., Dellatolas, G., Bancaud, J., & Talairach, J. (1988). Hemispheric lateralization of motor and speech functions after early brain lesion: Study of 73 epileptic patients with intracarotid Amytal test. *Neuropsychologia, 26*, 167-172.

Roberts, L. (1969). Aphasia, apraxia and agnosia in abnormal states of cerebral dominance. In P. J. Vinken & G. W. Bruyn (Eds.), *Handbook of clinical neurology: Vol. 4* (pp. 312-326). Amsterdam: North Holland.

Robinson, R. G., Kubos, K. L., Starr, L. B., Rao, K., & Price, T. R. (1984). Mood disorders in stroke patients: Importance of location of lesion. *Brain, 107*, 81-93.

Robinson, R. G., Starr, L. B., Kubos, K. L., & Price, T. R. (1983). Two-year longitudinal study of post-stroke mood disorders: findings during the initial evaluation. *Stroke, 14,* 736-741.

Robinson, R. G., Starr, L. B., Lipsey, J. R., Rao, K., & Price, T. R. (1985). A two-year longitudinal study of post-stroke mood disorders: In-hospital prognostic factors associated with six-month outcome. *Journal of Nervous and Mental Disease, 173,* 221-276.

Rosadini, G., & Rossi, G. F. (1967). On the suggested cerebral dominance for consciousness. *Brain, 90,* 101-112.

Rosen, A. D., Gur, R. C., Sussman, M., Gur, R. E., & Hurtig, H. (1982). Hemispheric asymmetry in the control of heart rate. *Abstracts of the Society of Neuroscience, 8,* 917.

Rosenbaum, T., DeToledo, J., Smith, D. B., Kramer, R. E., Stanulis, R. G., & Kennedy, R. M. (1989). Preoperative assessment of language laterality is necessary in all epilepsy surgery candidates: A case report. *Epilepsia, 30,* 712. (Abstract).

Ross, E. D., Edmondson, J. A., Seibert, G. B., & Homan, R. W. (1988). Acoustic analysis of affective prosody during right-sided Wada test: A within-subjects verification of the right hemisphere's role in language. *Brain and Language, 33,* 128-145.

Rossi, G. F., & Rosadini, G. (1967). Experimental analysis of cerebral dominance in man. In C. H. Millikan & F. L. Darley (Eds.), *Brain mechanisms underlying speech and language* (pp. 167-175). New York, Grune & Stratton.

Rouleau, I., Labrecque, R., Saint-Hilaire, J. M., Cardu, B., & Giard, N. (1989). Short-term and long-term memory deficit following intracarotid Amytal injection: Further support for the memory consolidation hypothesis. *Brain and Cognition, 11,* 167-185.

Rourke, B. P., & Brown, G. G. (1986). Clinical neuropsychology and behavioral neurology: Similarities and differences. In S. B. Filskov & T. J. Boll (Eds.), *Handbook of clinical neuropsychology: Vol. 2* (pp. 3-18). New York: John Wiley & Sons.

Sackeim, H. A., Greenberg, M. A., Weiman, A. L., Gur, R. C., Hungerbuhler, J. P., & Geschwind, N. (1982). Hemispheric asymmetry in the expression of positive and negative emotions. *Archives of Neurology, 39,* 210-218.

Säisä, J., Silfvenius, H., & Christianson, S. A. (1990). Visual half-field testing for defining cerebral hemisphere speech laterality. *Acta Neurologica Scandinavica, 82*, 346-349.

Sandson, J., & Albert, M. L. (1987). Perseveration in behavioral neurology. *Neurology, 37*, 1736-1741.

Sass, K. J., Lencz, T., Westerveld, M., Novelly, R. A., Spencer, D. D., & Kim, J. H. (1991). The neural substrate of memory impairment demonstrated by the intracarotid amobarbital procedure. *Archives of Neurology, 48*, 48-52.

Satz, P., & Fletcher, J. M. (1987). *Florida kindergarten screen battery.* Odessa, FL: Psychological Assessment Resources.

Satz, P., Orsini, D. L., Saslow, E., & Henry, R. (1985). The pathological left-handedness syndrome. *Brain and Cognition, 4*, 27-46.

Satz, P., Strauss, E., Wada, J., & Orsini, D. L. (1988). Some correlates of intra- and interhemispheric speech organization after left focal brain injury. *Neuropsychologia, 26*, 345-350.

Schwartz, A. S., Marchok, P. L., Kreinick, C. J., & Flynn, R. E. (1979). The asymmetric lateralization of tactile extinction in patients with unilateral cerebral dysfunction. *Brain, 102*, 669-684.

Schweiger, A., Wechsler, A. F., & Mazziotta, J. C. (1987) Metabolic correlates of linguistic functions in a patient with crossed aphasia: A case study. *Aphasiology, 1*, 415-421.

Scoville, W. B. (1954). The limbic lobe in man. *Journal of Neurosurgery, 11*, 64-66.

Scoville, W., & Milner, B. (1957). Loss of recent memory after bilateral hippocampal lesions. *Journal of Neurology, Neurosurgery, and Psychiatry, 20*, 11-21.

Serafetinides, E. A. (1966). Auditory recall and visual recognition following intracarotid sodium Amytal injections. *Cortex, 2*, 367-372.

Serafetinides, E. A., Driver, M. V., & Hoare, R. D. (1965). EEG patterns induced by intracarotid injection of sodium amytal. *EEG Clinical Neurology, 18*, 170-175.

Serafetinides, E. A., & Falconer, M. A. (1962). Some observations on memory impairment after temporal lobectomy for epilepsy. *Journal of Neurology, Neurosurgery, and Psychiatry, 25*, 251-255.

Serafetinides, E. A., Hoare, R. D., & Driver, M. V. (1964). A modification of the intracarotid amylobarbitone test: Findings about speech and consciousness. *The Lancet, 1*, 249-250.

Serafetinides, E. A., Hoare, R. D., & Driver, M. V. (1965). Intracarotid sodium amylobarbitone and cerebral dominance for speech and consciousness. *Brain, 88*, 107-130.

Silberman, E. K., & Weingartner, H. (1986). Hemispheric lateralization of functions related to emotion. *Brain and Cognition, 5*, 322-353.

Silfvenius, H., & Blom, S. (1984). Results from intracarotid Amytal tests in epileptic patients. *Acta Neurologica Scandinavica, 69*(Suppl. 99), 77-78.

Silfvenius, H., Blom, S., Nilsson, L., & Christianson, S. (1984). Observations on verbal, pictorial and stereognostic memory in epileptic patients during intracarotid Amytal testing. *Acta Neurologica Scandinavica, 69*(Suppl. 99), 57-75.

Snyder, P. J., & Novelly, R. A. (1991). An international survey of the administration and interpretation of the ISA procedure. (Unpublished manuscript).

Snyder, P. J., Novelly, R. A., & Harris, L. J. (1990). Mixed speech dominance in the intracarotid sodium Amytal procedure: Validity and criteria issues. *Journal of Clinical and Experimental Neuropsychology, 12*, 629-643.

Spiers, P. A., Schomer, D. L., Blume, H. W., Kleefield, J., O'Reilly, G., Weintraub, S., Osborne-Shaefer, P., & Mesulam, M-M. (1990). Visual neglect during intracarotid amobarbital testing. *Neurology, 40*, 1600-1606.

Strauss, E., Gaddes, W. H., & Wada, J. (1987). Performance on a free-recall verbal dichotic listening task and cerebral dominance determined by the carotid amytal test. *Neuropsychologia, 25*, 747-753.

Strauss, E., Satz, P., & Wada, J. (1990). An examination of the crowding hypothesis in epileptic patients who have undergone the carotid amytal test. *Neuropsychologia, 28*, 1221-1227.

Strauss, E., & Wada, J. (1983). Lateral preferences and cerebral speech dominance. *Cortex*, *19*, 165-177.

Strauss, E. & Wada, J. (1988). Hand preference and proficiency and cerebral speech dominance determined by the carotid amytal test. *Journal of Clinical and Experimental Neuropsychology*, *10*, 169-174.

Strauss, E., Wada, J., & Kosaka, B. (1985), Visual laterality effects and cerebral speech dominance determined by the carotid amytal test. *Neuropsychologia*, *23*, 567-570.

Tengesdal, M. (1963). Experiences with intracarotid injections of sodium Amytal: A preliminary report. *Acta Neurologica Scandanavia*, *39*(Suppl 4), 329-343.

Terzian, H. (1964). Behavioural and EEG effects of intracarotid sodium Amytal injection. *Acta Neurochirurgica*, *12*, 230-239.

Terzian, H., & Cecotto, C. (1959). Determinazione e studio della dominanzaemisferica mediante iniezione intracarotide di amytal sodico nell'uomo: I. Modificazioni cliniche. *Bollettino Societa Italiana di Biologia Sperimentale*, *35*, 1623-1626.

Teuber, H. L. (1974). Why two brains? In F. O. Schmitt & F. G. Worden (Eds.), *The Neurosciences: Third Study Program* (pp. 71-74). Cambridge: MIT Press.

Tucker, D. M. (1981). Lateral brain function, emotion, and conceptualization. *Psychological Bulletin*, *89*, 19-46.

Tucker, D. M. & Frederick, S. L. (1989). Emotion and brain lateralization. In H. Wagner & T. Manstead (Eds.), *Handbook of psychophysiology: Emotion and social behavior* (pp. 27-70). New York: John Wiley.

Tucker, D. M., Watson, R. T., & Heilman, K. M. (1977). Discrimination and evocation of affectively intoned speech in patients with right parietal disease. *Neurology*, *27*, 947-950.

Victor, M., & Agamanolis, D. (1990). Amnesia due to lesions confined to the hippocampus: A clinical-pathologic study. *Journal of Cognitive Neuroscience*, *2*, 246-257.

Victor, M., Herman, K., & White, E. E. (1959). A psychological study of Wernicke-Korsakoff syndrome. *Quarterly Journal of Studies on Alcohol*, *20*, 467-479.

Wada, J., & Rasmussen, T. (1960). Intracarotid injection of sodium Amytal for the lateralization of cerebral speech dominance: Experimental and clinical observations. *Journal of Neurosurgery, 17,* 266-282.

Walker, A. E. (1957). Recent memory impairment in unilateral temporal lesions. *Archives of Neurology and Psychiatry, 78,* 543-552.

Walker, J. A., & Laxer, K. D. (1989). Comparison of recall and recognition memory in the Wada test. *Epilepsia, 30,* 712-713. (Abstract).

Watson, R. T., Valenstein, E., & Heilman, K. M. (1981). Thalamic neglect: Possible role of the medial thalamus and nucleus reticularis in behavior. *Archives of Neurology, 38,* 501-506.

Weintraub, S., & Mesulam, M-M. (1985). Mental state assessment of young and elderly adults in behavioral neurology. In M-M. Mesulam (Ed.), *Principles of behavioral neurology* (pp. 71-124). Philadelphia: F. A. Davis.

Weintraub, S., & Mesulam, M-M. (1987). Right cerebral dominance in spatial attention: Further evidence based on ipsilateral neglect. *Archives of Neurology, 44,* 621-625.

Weintraub, S., & Mesulam, M-M. (1988). Visual hemispatial inattention: Stimulus parameters and exploratory stategies. *Journal of Neurology, Neurosurgery, and Psychiatry, 51,* 1481-1488.

Werman, R., Anderson, P. J., & Christoff, N. (1959). Electroencephalographic changes with intracarotid megimide and amytal in man. *EEG Clinical Neurology, 11,* 267-274.

Werman, R., Christoff, N., & Anderson, P. J. (1958). Behavioral and EEG changes induced by intracarotid drugs. *Transactions of the American Neurological Association, 83,* 74-76.

Werman, R., Christoff, N., & Anderson, P. J. (1959). Neurological changes with intracarotid Amytal and Megimide in man. *Journal of Neurology, Neurosurgery, and Psychiatry, 22,* 333-337.

Williams, J. J., Little, M. M., & Klein, K. (1986). Depression and hemispheric site of cerebral vascular accident. *Archives of Clinical Neuropsychology, 1,* 393-398.

Wolman, B. B. (Ed.). (1973). *Dictionary of behavioral science.* New York: Van Nostrand Reinhold.

Woods, R. P., Dodrill, C. B., & Ojemann, G. A. (1988). Brain injury, handedness, and speech lateralization in a series of amobarbital studies. *Annals of Neurology, 23*, 510-518.

Woods, B. T., Schoene, W., & Kneisley, L. (1982). Are hippocampal lesions sufficient to cause lasting amnesia? *Journal of Neurology, Neurosurgery, and Psychiatry, 45*, 243-247.

Wyllie, E., Lüders, H., Murphy, D., Morris III, H., Dinner, D., Lesser, R., Godoy, J., Kotagal., & Kanner, A. (1990). Intracarotid amobarbital (Wada) test for language dominance: Correlation with results of cortical stimulation. *Epilepsia, 31*, 156-161.

Yokoyama, K., Jennings, R., Ackles, P., Hood, P., & Boller, F. (1987). Lack of heart rate changes during an attention-demanding task after right hemisphere lesions. *Neurology, 37*, 624-630.

Zamrini, E. Y., Meador, K. J., Loring, D. W., Nichols, F. T., Lee, G. P., Figueroa, R. E., & Thompson, W. O. (1990). Unilateral cerebral inactivation produces differential left/right heart rate responses. *Neurology, 40*, 1408-1411.

Zatorre, R. J. (1989). Perceptual asymmetry on the dichotic fused words test and cerebral speech lateralization determined by the carotid sodium Amytal test. *Neuropsychologia, 27*, 1207-1219.

Zimmerberg, B., Glick, S. D., & Jerussi, T. P. (1974). Neurochemical correlate of a spatial preference in rats. *Science, 185*, 623-625.

Zoccolotti, P., Caltagirone, C., Benedetti, N., & Gainotti, G. (1986). Perturbations des résponses végétatives aux stimuli émotionels au cours des lésions hémisphèriques unilaterales. *Encephale, 12*, 263-268.

Zola-Morgan, S., Squire, L. R., & Amara, D. G. (1986). Human amnesia and the medial temporal region: Enduring memory impairment following a bilateral lesion limited to field CA1 of the hippocampus. *The Journal of Neuroscience, 6*, 2950-2967.

Senior Author Index